D0143281

Explosives Identification Guide,
2nd Edition

Mike Pickett
Associate Professor Emeritus
San Antonio College

THOMSON

DELMAR LEARNING

Australia Canada Mexico Singapore Spain United Kingdom United States

THOMSON

DELMAR LEARNING

™

Explosives Identification Guide, 2E

Mike Pickett

Vice President, Technology and Trades SBU:
Alar Elken

Editorial Director:
Sandy Clark

Acquisitions Editor:
Alison Weintraub

Development Editor:
Jennifer Thompson

Marketing Director:
Dave Garza

Channel Manager:
Erin Coffin

Marketing Coordinator:
Penny Crosby

Production Director:
Mary Ellen Black

Production Manager:
Larry Main

Production Editor:
Thomas Stover

Editorial Assistant:
Stacey Wiktorek

Cover Design:
Thomas Stover

Library of Congress Cataloging-in-Publication Data:
Card Number: [Number]

ISBN: 1401878210

NOTICE TO THE READER

DEDICATION

Explosives Identification Guide is dedicated to all emergency personnel who put their lives on the line daily and to explosive ordnance disposal personnel who pit their skills against these explosives. Most of all, this book is dedicated to Detective Bob Ellis, who was in charge of the San Antonio bomb squad. Bob was not only a friend but my mentor, and he gave his time freely to train many police officers and firefighters. He defeated many a bomb but could not defeat cancer. This book is for you, Bob.

I would also like to dedicate this book to a friend and fallen brother in blue, Chief Ray Downey, who was killed at the command post on 9-11 and to all first responders worldwide who have given their lives for the safety of mankind.

CONTENTS

Foreword   vii

Preface   ix

Introduction   xi

    Why Emergency Personnel Should Know
        About Explosives   xi
    Safety Procedures   xi

Chapter 1 **Identifying Some Common Commercial Explosives** **1**
    Blasting Caps   1
    Safety Fuse   2
    Detonating Cord   3
    Blasting Agents   3
    Boosters   3
    Dynamites   4
        Nitroglycerin Dynamite   4
        Ammonium Dynamite   4
    Slurries or Gels   5
    Black and Smokeless Powders   5
    Linear Shape Chargers   5
    Conclusion   6

Chapter 2 **Military Ordnance: Deadly Souvenirs,**
          Unexploded Ordnance   **7**
    Color Coding   8
    Fuzes   8
    Grenades   9
    Ammunition (Shells or Projectiles)   10
    Mortar Shells   11
    Rockets and Missiles   11
    Mines   12
    Submunitions   13
    Cartridge-Activated Devices   14
    Air-Dropped or Gravity Bombs   14
        Smart Bombs   15
        Fragmentation Bombs   15
    Bulk Explosives   15
        C-3 and C-4 Explosives   16

	Detonating Cord	16
	Incendiaries	**16**
	Simulators	**17**
	Chemical Ordnance	**18**
	Nuclear Weapons	**18**
	Conclusion	**19**
Chapter 3	**Bomb Threats and Bomb Search Procedures**	**21**
	Triggering Mechanisms of Improvised Explosive Devices	**21**
	Booby Traps and Antipersonnel Devices	**26**
	Bomb Threats	**27**
	Motives of Bombers	27
	Mental Disorders	29
	Phone Threats	29
	Mail Threats	31
	Bomb Searches	**31**
	Planning	31
	Equipment	32
	General Search Procedures	33
	Conducting a Room Search	36
	Evacuation	36
	Conclusion	**38**
Chapter 4	**Weapons of Mass Destruction**	**39**
	Nuclear Weapons	**40**
	Weapons	40
	Radiological Dispersal Device	41
	Biological Weapons	**44**
	Bacteria	44
	Viruses	45
	Toxins	46
	Chemical Agents	46
	Choking Agents	48
	Blood Agents	48
	Blister Agents	49
	Conclusion	50
Appendix A	**Color Insert Captions**	**51**
Appendix B	**Color Coding**	**57**
Appendix C	**Checklist**	**59**
Appendix D	**Glossary**	**61**

FOREWORD

I congratulate Mike Pickett on the 2nd edition of his definitive book on identifying explosives. It is an important addition to the literature and learning for fire and emergency personnel in the new millennium, and a key reference book for those who have to deal on a daily basis with the increasing threats around our world.

There is increasing attention to the areas of fire and rescue, emergency management, emergency medical services, law enforcement, homeland security, national security, and related areas since September 11, 2001, and I see our profound and professional efforts in these areas as the best way to honor those who so valiantly gave their lives in those attacks to save others.

More and more books are appearing on terrorism and related topics. so there is no dearth of books on similar topics. However, this book stands out because it drills down to a very technical and specific level, with tools and techniques which empower emergency responders to protect themselves, the victims, and their communities.

Not only are there lots of new books, there is a new think tank, new higher education degree programs, new areas of research and development; however, the traditional professions and disciplines provide the foundation, and the studies of hazardous materials and explosives provide the basic handles on those dangerous materials most often used by terrorists.

Much of the training offered to emergency personnel is at the awareness level and the strength of this book is that it takes the reader/student to the next level of knowledge, professionalism, and expertise. It is, in essence, a how-to book on the hottest topic of the day, presented with the expertise developed through hard-earned experience of a world-renowned authority. It will be a book you will appreciate, learn from, and want to keep around for ready reference.

Around the world, we are spending about $100 billion a year on emergency and disasters, so it is a big business and a large component of the world economy. Increasingly, explosives play a significant role in these incidents in both war and peace settings.

We need to work as hard at prevention and preparedness, as the terrorists do at destruction and desecration. As Winston Churchill once said, "This is not the end. This is not even the beginning of the end. This is just the end of the beginning." I believe that is about where we are in emergency preparedness in this country. This book can take us a long way toward the next level.

KAY C. GOSS, CEM(r)
Senior Advisor for Homeland Security and Emergency Management, EDS

Chair, Education and Training Committee, International Association of Emergency Managers and former Associate Director of FEMA

PREFACE

■ INTENDED AUDIENCE

I wrote the *Explosives Identification Guide* because of the lack of knowledge about explosives available to emergency services. My six years of working with explosives while in the U.S. Air Force and 30 years as a firefighter have borne this out time and time again.

This book, therefore, is intended to be a helpful reference guide for any first responders – whether they be law enforcement, firefighter, security personnel or EMS. This guide is designed to assist them in identifying the thousands of different explosives they may encounter in their duties of protecting the citizens under their care. It is for all first responders who, at 3 am in the morning in the pouring rain, are trying to determine what they are looking at, and need this information for effective, and safe, decision-making.

■ HOW TO USE THIS BOOK

This book is a handy reference to use when responding to incidents involving potential explosive materials. It is divided into four easy-to-access chapters:

Chapter One: Commercial Explosives
Chapter Two: Military Ordnance
Chapter Three: Bomb Threats & Search Procedures
Chapter Four: Weapons of Mass Destruction

Each chapter categorizes the types of explosives, providing descriptions and photos for quick identification. Practical, straight-forward information on search procedures allows for quick decision-making, and valuable information on WMD or CBRNE, ensures the book is up to date, and informative, regarding the latest threats to our society today.

■ FEATURES OF THIS TEXT

A **Full – Color Insert** provides photos and short descriptors of all the explosives discussed in the book – allowing for a quick identification while responding to incidents involving explosive materials.

A **Safety Checklist** outlining Standard Operating Guidelines (SOG) emphasizes the need for safety to ensure effective and successful response to an explosives incident.

A **Glossary** provides a list of acronyms, critical terminology, as well as an interpretation of slang for a quick study.

■ NEW TO THIS EDITION

When I wrote the first edition of the book, I had a personal reason to focus on terrorism, as the World Trade Center bombing in February 1993 involved some New York City firefighters that I had worked with, and the Oklahoma City bombing involved some of my family in Oklahoma City. For this reason, I choose to focus on the day-to-day routine encounters that a first responder may face in regards to explosives. Since then, and in light of the events of 9/11, world events have altered my approach. Therefore, the second edition now focuses on terrorist activities on a larger scale that may or may not involve explosives.

A new chapter on Weapons of Mass Destruction (WMD), now known as CBRNE (Chemical, Biological, Radiological, Nuclear, and Explosive) brings into focus the latest threats to society – explaining each type and the recognizable symptoms for effectively dealing with each situation.

A Glossary offers a reference list for commonly used terminology in responding to explosives and WMD incidents.

Please visit us at *www.firescience.com* for more information on other emergency response titles offered by Delmar Learning!

■ ABOUT THE AUTHOR

Mike Pickett has over thirty-seven years experience as a firefighter, including ten years as an EMT. He was the director of the San Antonio Regional Fire Academy and Fire Science Program for twenty-five years. He is currently an associate professor emeritus and teaches explosive identification and counter-terrorism to audiences across the country.

He spent six years in the USAF working on all types of ordnance, including nuclear and chemical weapons. He is a member of the National Explosives Ordnance Disposal Association and the USAF's Explosives Ordance Disposal MasterBlasters Association, Inc. He holds a master's degree in Urban Planning and a bachelor's degree in Nuclear Chemistry.

His Web site is *www.pickettsprimer.com*.

■ ACKNOWLEDGEMENTS

The author and Publisher would like to thank the following individuals whose careful review and contribution to the 2nd Edition have helped us to create a technically accurate and timely book:

William Hand
Houston Fire Department
Hazmat Response Team
Houston, TX

Danny Peterson
Professor
Arizona State University – East
Mesa, AZ

Glen Rudner
Hazmat Officer
Virginia Department of Emergency Management
Dumfries, VA

■ AUTHOR'S ACKNOWLEDGEMENTS

Anne Arreaga

Captain Jeffery Ford of the 137th EOD, Fort Sam, Houston, Texas

SGM R. K. Harrison, USAOMMCS, Redstone Arsenal, Alabama

MSgt Henry, 96 CEG/CED, Eglin Air Force Base, Florida

MSgt Gerald Jarrell and the men of the 37th EOD, Lackland Air Force Base, Texas

Robert Ojeda, Fire Chief, San Antonio Fire Department

The guys of Engine 17 and Truck 17, some of the best and worst cooks in the world

INTRODUCTION

Explosives are a necessary part of our world. They help us to extract vital metals and materials from the ground and to remove old, unused buildings. We use them to blast through rock to build roads, tunnels, and railroads, and to lay foundations for new homes. Explosives are used for hundreds of important jobs and services that make our lives easier and safer.

Today's news is full of terrible stories of bombings and terrorists, and of people being killed and property being destroyed because of explosives. Unfortunately, it is the misuse of these explosives that causes the stories that make the headlines. Our first responders, such as police officers, firefighters, and emergency medical personnel, need to learn more about explosives in order to respond safely to such incidents.

Emergency personnel need to learn to recognize and identify explosives for many reasons. They can encounter them in their normal duties. I have been on duty when citizens have brought into the fire station, dynamite that their children had found at a construction site and another time, a hand grenade their father had kept as a war souvenir. I found a military flare after we extinguished a garage fire. First responders can also encounter explosives being manufactured, transported and used in many different ways. Many formerly used defense sites (FUDS) or old military installations are being closed down across the nation. There are some 9600 FUDS, many of which were old bombing and firing ranges and contains unexploded ordnance (UXO). Some day, this old ordnance may surface.

■ SAFETY PROCEDURES

When responding to an explosive incident, several safety procedures should be followed. Even if an explosive has detonated, unexploded explosives or ordnance may still be laying around.

Each department or organization should have a standard operating procedure (SOP) for explosives, before one is needed.

❐ Do not use two-way radios, radar, or television transmitting devices within 1,000 feet. This includes the mobile data terminals (MDTs) in the apparatus and cellular phones. The electromagnetic radiation (EMR) given off by these devices may detonate the item. See *SLP 20 Safety Guide for the Prevention of Radio Frequency Radiation Hazards in the Use of Commercial Detonators (Blasting Caps)*, Institute of the Makers of Explosives (IME), July 2001.

❐ Notify the proper authorities, depending on the jurisdiction and the situation. The fire department, police department, sheriff, military explosive ordnance disposal units, or Bureau of Alcohol, Tobacco and Firearms (ATF) personnel may be involved. Check your SOP on explosive incidents.

❏ Clear and control the area as you would during any hazardous materials incident. The size and type of explosive, terrain, shielding, and other factors will determine the size of the area to be controlled.

❏ Stage emergency medical service (EMS), fire, and police units outside the control point. Emergency units are of little use if they are destroyed in a blast.

❏ *Do not approach* the suspected explosive because it may have motion-sensitive or acoustic fuses that function once they sense a target. Use binoculars to observe the area.

❏ Reduce the potential effects of a blast and flying shrapnel by opening doors and windows and by placing emergency vehicles in the path of the blast wave to act as a shield.

■ 1 ■

IDENTIFYING SOME COMMON COMMERCIAL EXPLOSIVES

■ BLASTING CAPS

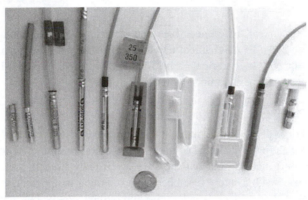

Figure 1–1 Nonelectric blasting caps.

Among the most commonly found explosives are blasting caps used to set off main charges. These small, thin-walled cylinders are silver or copper in color. They are $1/4$ to $1/2$ inch in diameter and from 2 to 6 inches in length. Blasting caps contain a small amount of a sensitive but powerful explosive. As with all explosives, they should not be handled. It has been said that even the heat of a person's hand can set off a blasting cap.

Blasting caps come in two types: mechanical and electrical. Mechanical or non-electrical caps are set off by a safety fuse inserted into the end. New types of caps have the fuse already inserted called shock tube whereas older types blasting caps have an open end for the fuse to be inserted. Misfires are common with these caps, particularly with older types. The black powder core of the safety fuse often gets wet, and unexploded caps are often found scattered around blasting sites. Because of this, often two caps per blast are used to ensure detonation.

Electrical caps are exploded by the heat between two electrical wires that extend out the end. These wires come twisted together, or shunted, at the end so that the cap is not set off by a stray electrical current. Such a current can come from static electricity on a person's body as well as from radio, television, or radar transmissions, known as electromagnetic radiation, or EMR.

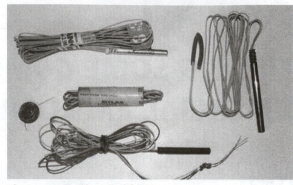

Figure 1–2 Electric blasting caps.

There are few physical differences between electrical and mechanical, or non-electrical, blasting caps. In mechanical caps, there are normally either no wires or only one tube at the end. In electrical caps, there are two wires at the end of the cap. All caps, regardless of type, should be left alone.

■ SAFETY FUSE

A safety fuse is a time-delay device much like a firecracker fuse. It has a black powder core covered by a waterproof jacket of solid or striped colors of orange, white, or black. These fuses will burn underwater, but they are not usually used in wet conditions. Caution must be used with a safety fuse because it is similar in appearance to detonating cord. Older fuses had a black powder core and burned slowly, but newer blasting caps have a shock tube instead of a safety fuse.. This tiny shock tube detonates instead of burning. They must be set off by shock, not flame, by special initiators.

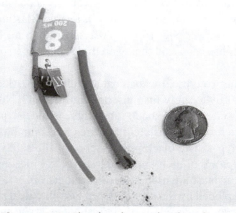

Figure 1–3 Shock tube and safety fuse.

■ DETONATING CORD

Detonating cord, commonly known as det cord or Prima Cord®, is similar in appearance to safety fuses. It comes in various colors of spools holding up to 2,000 feet of cord. U.S. military det cord is usually in olive drab color on the outside, but it can come in other colors, including bright colors. All det cord has a white core of a powerful explosive that detonates at about 4 miles per second. This cord is normally used to set off multiple charges simultaneously, but there are many other uses for it, both legal and illegal.

Figure 1–4 Detonating cord.

BLASTING AGENTS

Blasting agents are a combination of a fuel and oxidizers such as fuel oil and ammonium nitrate fertilizer (NH_4NO_3), sometimes known as nitro carbo nitrate or ammonium nitrate fuel oil (ANFO). As a fertilizer, NH_4NO_3 is relatively stable and burns quietly unless it is confined or contaminated. It can be purchased cheaply by the ton as a fertilizer and is thus often used in bombings. NH_4NO_3 in its pure form is a white grain similar to table salt, and it will absorb moisture and harden. As a blasting agent, it can range in color from light brown to bright pink due to the added fuel, which also gives it a strong odor. Blasting agents require a booster charge of high explosive since unconfined, they cannot be set off by a blasting cap. Blasting agents were used in both the New York City World Trade Center and the Oklahoma City bombings.

■ BOOSTERS

Boosters are explosives in which blasting caps are placed to increase the power of the initiating charge. Some are cylinder shaped and are about the size of a penlight or marking pen. Others are about the size and shape of a soft drink can and are white. Boosters also come in a variety of other sizes and colors, but they always have a hole in which to insert the blasting cap.

Figure 1–5 Booster explosive for blasting agent.

■ DYNAMITES

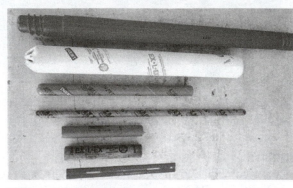

Figure 1–6 Dynamite.

There are two basic types of dynamite, nitroglycerin and ammonium dynamites. Dynamite comes in a variety of shapes, sizes, and colors. Cartridges, or sticks, are typically from $1^{1}/_{8}$ to 3 inches in diameter and from 8 to 24 inches in length, but they can be even larger. Cartridges range in color from light brown to orange-red in fiberboard or waxed-paper sticks. They are typically shipped in 50-pound cardboard boxes.

Nitroglycerin Dynamite

Nitroglycerine dynamite is nitroglycerin (nitro) mixed with sawdust or a gray clay-like filler. The nitro settles at the bottom of the sticks, oozes out of them, and crystallizes. This process, known as exudation or sweating, can cause the sticks to adhere to each other. Pulling them apart can cause detonation. Old dynamite can be extremely dangerous even to trained personnel due to its instability.

Ammonium Dynamite

Ammonium dynamite has a high ammonium nitrate content. The color and size of these sticks are similar to nitro dynamite. For our purposes, dynamite is dynamite.

■ SLURRIES OR GELS

Slurries or gels are among the newer explosives. Slurries are liquid explosives, and gels are semiliquid explosives. Either can be poured and are stored in large sausage-shaped tubes.

Figure 1–7 Slurries.

■ BLACK AND SMOKELESS POWDERS

Figure 1–8 Black and smokeless powders.

Black and smokeless powders are common over-the-counter low explosives used by gun enthusiasts to reload ammunition or used in old powder-loading guns. Unconfined, they will deflagrate (burn intensely) but will detonate when confined. They are often used in homemade bombs.

There are many other varieties of commercial explosives, including flexible plastic explosives such as Flex-X™, and linear shape charges. Linear shape charges are usually used to cut beams and columns, and S.W.A.T. teams use them for forcible entry, Figure 1–9.

Figure 1–9 Linear shape charge.

■ CONCLUSION

Commercial explosives serve a variety of useful purposes, but may be encountered by first responder, in the performance of their duties. A good source of information is the Institute of the Makers of Explosives (IME) 1120 19th Street, N.W., Suite 310, Washington D.C., 20036-3605, phone 202-429-9280 or the local ATF office.

■ 2 ■
MILITARY ORDNANCE: DEADLY SOUVENIRS, UNEXPLODED ORDNANCE

Discussion of all varieties of current military ordnance would be impossible here; there are thousands of types of ordnance. To complicate matters, older ordnance remains from wars as far back as the Civil War. Many formerly used defense sites (FUDS) or old military installations are being closed down across the nation. There are some 9,600 FUDS, many of which were old bombing and firing ranges and contain unexploded ordnance (UXO). Many housing developments are being built over or next to these sites. Some are being turned into parks. Some day, this old ornance may surface as the soil around them erodes, and when it does, the first responder will be called. With the downsizing of the military and the closing of bases, this problem will increase. Citizens bring current and old ordnance into police and fire stations. Two hand grenades were recently deposited at fire stations in San Antonio, Texas. Some military personnel bring ordnance home as souvenirs; they consider such devices harmless because they have failed to detonate for various reasons. Military ordnance can still explode and, unfortunately, some of these items find their way into homes and into the hands of children.

This chapter provides an overview of military ordnance. It is not intended to be comprehensive or technical, but to give a general representation of the many types of ordnance. Certain types have similar shape and differ only in size. Readers should be aware that the ordnance colors and marking described in this book are among those frequently used, but variations are common. When reporting an item, one should give the following information *without* moving or touching the item:

- Size: length, width (or diameter)
- Color: main body, any painted bands or stripes
- Markings: stamped or painted on the surface
- Material: plastic, wood, metal, and so forth
- Fuze: Does it have a fuze?

■ COLOR CODING

The color of the body of the ordnance and the markings can be helpful, but they should not be totally trusted for the following reasons:

- Different countries have different color codes.
- Colors may have been changed from the original, especially with war souvenirs and training items.
- Discoloring can occur through sun bleaching, rusting, and so forth.
- Booby-trapped items may be colored to blend in with the surroundings.
- Color coding has changed over the years.

As a ground rule, U.S. coloring is as follows:

Yellow: high explosive (HE)

Black: armor piercing (AP)

Blue: training items or items with a hole or holes drilled

Brown: propellant

Gray: chemical such as a riot grenade

See Appendix B for more detail on color coding.

■ FUZES

Figure 2–1 A mixture of bomb and projectile fuzes.

Fuzes are used to initiate the explosive in any warhead, whether a bomb, projectile, rocket, or any piece of ordnance. The action of a fuze can be point detonated (PD) by impact, proximity, or delay. PD fuzes act on the impact with the target, but fraction-of-a-second delays can be built in. A proximity fuze (variable time [VT]) detonates when the warhead is at a certain distance from the target. Walking near a proximity fuze can cause detonation. Delay fuzes can function seconds, minutes, or days after the device is delivered. Delay fuzing can be chemical, electrical, or mechanical, and, as with proximity fuzes, should not be approached! There is no way that untrained personnel can know what type of fuze is present, so any item must be considered to be the most dangerous type. Stay well clear of *any* ordnance. Use binoculars to observe, as you would in any hazardous incident. The fuzes may not always be visible when installed in the warhead, or they may be buried in the ground. With many types of ordnance, the fuzes in transport are stored separately from the warhead. The fuzes are not armed until the ordnance is dropped from an aircraft, launched, or fired. The centrifugal force from the spin of the projectile or the inertia of the projectile leaving the gun barrel or mortar tube arms some fuzes.

Most nose bomb fuzes have vanes or impellers on the front that turn in the airstream to line up the firing pin once the bomb leaves the aircraft. An arming wire that runs through the vane prevents the vanes from turning until dropped. The arming wire slips through the vane and stays with the aircraft after the bomb is dropped, allowing the vane to turn and the fuze to arm. Tail fuzes do not have the vane on the fuze itself; instead, the fuze is linked to the vanes by either a flexible or a rigid shaft. Electrical fuzes use electrical power from the aircraft and are armed just before the bomb is released from the aircraft. Some have batteries built into them. Some VT and always-acting fuzes are electrically armed. Some fuzes such as the M 904 have a small window showing green for unarmed and red for armed. Bomb fuzes are normally not installed in a bomb until it is mated to the aircraft.

■ GRENADES

Figure 2–2 Hand grenades.

Probably the most common military ordnance is grenades. Some thrown grenades have a long handle, or spoon, placed on top of the fuze. The handle extends from the top down the side of the grenade, approximately three-quarters the length of its body. A cotter or safety pin with an O-shaped ring is inserted through the spoon and into the side of the square fuze on top of the grenade. The spoon holds the spring-loaded striker down and is locked by a safety pin. Pulling the O-shaped ring releases the spring-held fuze and spoon, arming the grenade and causing it to detonate within seconds. The time varies with the type and age of the grenade. Some newer models contain electrical fuzes.

The old U.S. Army fragmentation-type grenade is olive drab in color, measures $2^3/_8$ inches in diameter and $4^3/_4$ inches in length and weighs 21 ounces. Because it is shaped like a small pineapple, it is sometimes called the pineapple grenade. A square fuze is located on its top. Newer grenades are also olive drab in color, are the size and shape of baseballs, and have the same type of fuze on the top as older grenades.

Other types of grenades contain smoke, white phosphorus (WP), tear gas, or riot gas. These can be shaped like baseballs or soda cans and can come in a variety of colors, including gray and red. Nearly all grenades have identical-looking fuzes. Some rifle grenades can look like small bazookas or rocket shells. Some, like 40-mm rifle grenades, look like large, fat bullets. The distinctions between rocket-propelled grenades (RPGs) and other rockets such as the light antitank weapons (LAWs) are almost nil.

■ AMMUNITION (SHELLS OR PROJECTILES)

Figure 2–3 Artillery shells.

Ammunition, or ammo, ranges from arms as small as 5.56-mm up to 16-inch shells. Shells have the typical bullet shape with a brass-colored casing with a projectile, or warhead, on top. Some larger projectiles require the propellant to be separate from the warhead and to come in bags. Ammunition 20 mm and larger can have a warhead that contains a fuze and HEs; these are considered miniature bombs and are called shells or projectiles rather than bullets. Fuzes on the projectiles are activated either by the spinning of the shell or by the setback inertia, or G force, involved in firing. The fuzing can be impact (PD), time delay, or proximity (VT). A proximity or VT fuze can be detonated by the approach of a person or vehicle. Assume that all ordnance may have a VT fuze and stay clear!

Some shells have a brass-colored band near the end of the projectile that is scored, or rifled, in order to spin and stabilize the projectile when fired. It is often called a rotating band although it does not actually rotate. Rifling marks on a projectile indicate that it is probably armed. A lack of marks on rotating bands indicates that the device has not been fired and that the fuze is probably not armed if it is a U.S.-made round of ammunition. This may not be true of foreign-made rounds. Some projectiles may have fins for stabilizing. Projectiles, as well as rockets, mortars, and bombs, can contain HE fillers, chemicals, WP, submunitions, and pyrotechnics. Some shells nicknamed beehive rounds contain hundreds of tiny metal darts called flechettes.

■ MORTAR SHELLS

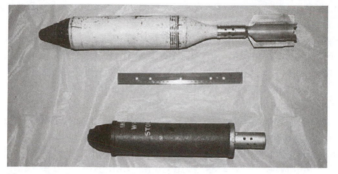

Figure 2–4 Mortar shells.

Mortar shells look much like projectile shells except that many have no brass casing or rotating band. Propellants are clipped or attached to the backs of the shells. U.S. mortars range in size from 60 mm to 4.2 inches in diameter and up to 26 inches in length. They weigh up to 27 pounds. Foreign-made, especially Soviet, mortar shells can be larger. The fuzing is the same as with projectiles. Some shells have fins for stabilizing. There is little indication that a mortar round has been fired or that a warhead is armed.

■ ROCKETS AND MISSILES

Figure 2–5 Missiles.

Rockets and missiles are devices that are self-propelled through the air to such targets as aircraft, personnel, or vehicles. They generally consist of a solid rocket motor and a warhead filled with HEs, chemicals, WP, submunitions, or illumination flares. Some rockets have the same type of fuzing as mortars and projectiles; this fuzing is located internally between the warhead and the rocket motor or in the nose of the warhead. As a general rule, the warhead is shipped and stored separately from its launcher, except for the smaller, shoulder-fired rockets, which are shipped in their tube or launcher. Some rockets are spin-stabilized but do not have a rotating band like a projectile or mortar. This spin is caused by fins (either fixed or folding) or by the exhaust of the motor.

LAW rockets, sometimes referred to as bazooka rounds or rockets, can be fired by individuals (shoulder-fired). Rockets can also be fired by aircraft, helicopters, or tanks or from stationary positions.

Guided missiles are similar in appearance to rockets except that they contain some sort of guidance system. Smaller missiles like the Dragon missile are wire-guided by the gunner: a wire is fed out of the back as the missile flies toward the target. The tube-launched, optically tracked, wire-guided (TOW) missile is similar. Some shoulder-fired missiles such as the Stinger are guided by infrared and home in on the hot exhaust of an aircraft. The air-intercept missile AIM-9 Sidewinder is similar in its guidance system. Other guidance systems include radar, laser, terrain recognition, and optical—that is, guided by a television camera (TV).

The fuzing and types of warheads are similar to rockets and projectiles. Missiles come in a variety of sizes and for various purposes:

Air-intercept missiles (AIMs) are designed to shoot down aircraft. Typically, they are launched from other aircraft. An example is the AIM-9 Sidewinder, which homes in on the exhaust of an aircraft.

Air-to-ground missiles (AGMs) are air launched on targets such as vehicles or buildings. An example is the AGM-88 high-speed antiradiation missile, (HARM), which is designed to attack radar sites.

Surface-to-air missiles (SAMs) are surface launched to shoot down aircraft. Russian SA-2s were used to shoot down U.S. aircraft in Vietnam, and one brought down a U.S. U-2 spy plane over Russia on May 1, 1960, with pilot Gary Powers in it.

Surface-to-surface missiles include the intercontinental ballistic missile (ICBM) and the intermediate-range ballistic missile (IRBM). These missiles are rather large and often contain a nuclear or chemical warhead. The Scud missiles launched by Iraq during the Gulf War were surface-to-surface missiles.

■ MINES

Figure 2–6 Land mines.

Mines are explosives designed to (1) destroy an enemy in place or (2) prevent or deny the enemy entry into an area. They are deployed in two ways. Scatterable mines are scattered over the surface by aircraft, helicopter, projectiles, and rockets and are visible on the surface. Placed mines are buried or hidden, usually in some sort of a pattern. Of all ordnance worldwide, mines are the leading cause of injury and death to civilians, even long after the conflict is over. The United Nations claims that more than 10,000 people are killed and thousands more are maimed each year. Many of these victims are children hurt by leftover mines.

Antipersonnel (APERS) mines are small and come in different shapes and materials, including metal, wood, and plastic. Some are designed as booby traps, some detonate when stepped on, and some are activated by trip wire. Antitank (AT) mines are larger than APERS mines and are usually pressure detonated, but they may also have magnetic, trembler, or acoustical fuzes. To prevent detection by metal detectors, many mines are nonmetallic. Some mines are buried on top of each other, making a booby trap for mine-clearing personnel or have antilift wires attached to detonate when lifted.

■ SUBMUNITIONS

Figure 2–7 U.S. submunitions.

Submunitions can be mines or small bomblets. They are called bomb live units (BLUs) and are air dropped from aircraft, helicopters, or projectiles. The dispensers are designated suspended underwing unit (SUU) when empty and cluster bomb units (CBUs) when filled with the BLUs. The type of BLU will determine the CBU designation. It is difficult if not impossible to determine the type of CBU from the exterior. There are over a hundred different U.S.-made types of CBUs. Many are still classified by the military. One of the first BLUs was the M 83, called the butterfly bomb; its vanes unfold in flight to slow its fall and to arm the bomblet, resembling a butterfly. A newer version is the BLU-3, which is about 3 inches in diameter and 5 inches long. It has a yellow body and metallic-colored fins that fold around the body until it is dropped. The fins of the BLU-3 fold out and up when it is dropped to rotate the bomblet, stabilizing and arming it. The BLU-7 is a bomblet used as an antitank weapon. When it lands, it can blow a hole through armor plate.

It is $2^3/_4$ inches in diameter, 9 inches long when closed, and 28 inches long when opened. In its open position, the BLU-7 has a ribbon parachute to stabilize it, slow its fall, and arm it. Americans dropped millions of BLU-3s and BLU-7s in Vietnam. Some submunitions act like mines and will blow up if pressure is applied, whereas others have antidisturbance fuzing with a self-destructing fuze that can function hours or days later as a backup. Some bomblets have a shape charge for penetrating hard targets. These usually have a parachute or ribbons to stabilize and aim the bomblet. Others have fins for the same purpose.

◼ CARTRIDGE-ACTIVATED DEVICES

There are many small cartridge-activated explosive devices on aircraft. Because of the small size of such devices, aircraft maintenance crews sometimes regard them as harmless. For example, a worker may place such a device in a pocket, take it home, and inadvertently leave it out in the open where a child might find it.

Impulse cartridges look like shotgun shells except they are usually metallic in color. Their size varies. They are used for the ejection of pylons (devices used to attach equipment to aircraft), launchers, racks, fuel tanks, missiles, conventional bombs, and special stores from the wings and fuselage of aircraft. Other cartridges are used for starting jet aircraft engines. Some act as guillotines for cutting various hoses or electrical lines. For example, many ejection seats in aircraft have projectile or rocket motors. A small triangle alongside the fuselage of an airplane, just below the canopy, indicates the presence of an ejection seat.

◼ AIR-DROPPED OR GRAVITY BOMBS

Encounters with air-dropped weapons are rare except in shipping accidents or air-craft crashes. General-purpose bombs, nicknamed "iron bombs," commonly come in 250-, 500-, 750-, and 2,000-pound classifications.

Older high-drag bombs are shaped like a fat bullet with box-shaped fins on the ends. They have one or more yellow bands on their nose and tail, which indicate an HE filler; the body or bomb casing is olive drab in color. Newer low-drag bombs are more streamlined, with conical-shaped fins. Some have a parachute or fins that act as speed brakes to slow the bomb, allowing the aircraft to escape before the bomb explodes. These bombs also have a yellow band on the nose, indicating an HE filler.

A gray body and red or green stripes indicate a chemical bomb. These bombs can be lethal while in storage or shipment if the filler is leaking.

The body of an air-dropped bomb has an opening in the nose and tail for the fuze and booster. The fuze slips inside a booster cup and is inserted into the bomb. The fuze contains small but powerful explosives on the end. The front has vanes, or impellers, that turn in the airstream, causing the bomb to be activated.

Most U.S.-made fuzes have a small window. A green color in the window indicates that the fuze is unarmed; red indicates that it is armed. Other fuzes include chemical delay, proximity (VT), and always-acting fuses. These are different in shape and may or may not have indicators to reveal whether or not they are armed.

Figure 2–8 Smart bomb.

As with all ordnance, such devices should be left alone and immediately reported to bomb-disposal professionals.

Smart Bombs

Figure 2–9 Fragmentation bomb.

The term *smart bomb* refers to a regular iron bomb but with an attached guidance system kit of some sort. This kit includes a guidance system and fins or wings for directional control. The hazards to personnel are the same as with regular "iron bombs".

Fragmentation Bombs

Fragmentation bombs (frags) are about 20 pounds or more in weight and have bodies spiral wound with thick steel wire. Frags use only nose fuzes. They are dropped in clusters so they are sometimes called cluster bombs.

■ BULK EXPLOSIVES

Trinitrotoluene (TNT) is the explosive by which all other explosives are measured. It is light yellow in color. The $1/4$-pound size comes in a round cardboard cartridge $1 1/2$ inches in diameter and $3 1/2$ inches in length. The $1/2$-pound size uses a rectangular cartridge $3^3/_4$ by $1^3/_4$ by $1^3/_4$ inches. The 1-pound size is 7 by $1^3/_4$ by $1^3/_4$ inches. TNT cartridges are olive drab in color with yellow markings and a blasting cap well in the end.

C-3 and C-4 Explosives

The C-3 explosive is a plastic explosive that can be molded like putty. It comes in clear plastic or olive drab–colored cartridges 12 by 2 by 2 inches, similar to TNT, and is yellow in color. The C-4 explosive is a newer version. It is pure white in color with 11 by 2 by 2 inch dark green cartridges. In addition, there are other types of military explosives, including shape charges, cratering charges, and Bangalore torpedoes.

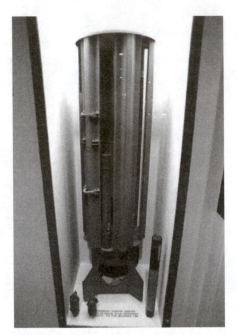

Figure 2–10 Incendiary bomb and cluster.

Detonating Cord

Military det cord comes in a variety of exterior colors, including white and olive drab.

■ INCENDIARIES

Incendiaries are devices that burn rather than explode. They are classified by the U.S. Department of Transportation as Class 1.4 explosives. They are approximately 22 inches in length with a 3-inch hexagonal-shaped diameter. Incendiary bombs were first developed in World War II for use in bombing cities. They were originally dropped from airplanes in clusters of 30 to 130 bombs at a time. These bombs break apart in the air, starting scattered fires over large areas. Because some contain magnesium and sometimes a bursting charge, it is useless and dangerous to try to extinguish them.

Another bomb developed in World War II was napalm. A soap compound was mixed with gasoline to make a sticky solution. This mixture was then poured into external fuel tanks with WP igniters and dropped from aircraft. Modern napalm bombs have changed little since their conception. Only a trained observer can distinguish them from regular aircraft fuel tanks. One bomb, a BLU-52 chemical bomb is a modified BLU-1/B incendiary bomb. A new class of bombs are called fuel air explosives (FAE). Some FAEs use propane or ethylene oxide as a fuel that creates a fuel mist and then ignites. Some FAEs are still classified, so information on them is not available.

Photoflash bombs are about 5 feet long and 10.6 inches in diameter. They contain about 75 pounds of photoflash powder and are used for night photography. They are quickly becoming obsolete.

Flares serve a variety of purposes. They illuminate battle scenes, help spot downed aircraft crew, and act as decoys for heat-seeking missiles. Flares differ from

Figure 2–11 Trip flare.

photoflash in that they burn in 6 to 60 seconds, whereas photoflashes burn instantaneously, like camera flashbulbs. Some common types of flares look like an aluminum tube, 10 to 12 inches in length and 2 inches in diameter and weigh from 1 to 1.5 pounds. Some flares contain expelling charges with parachutes. Some flares, such as the LUU-10, can be 4 inches in diameter and 30 inches long. Some trip flares look like small hand grenades with a similar fuze and operation.

■ SIMULATORS

Simulators are large firecrackers that simulate battle noises and effects. The M 116A1 hand grenade simulator is a cylindrical paper tube 4.3 inches in length and 2.18 inches in diameter with a 5- to 10-second fuse and igniter along the side. It is white with red tape holding the igniter to the body. The M 115A2, a projectile groundburst simulator, is similar but smaller; it is 7.13 inches long and 2.4 inches in diameter.

The M 21 artillery flash simulator consists of an outer plastic case 5.9 inches long and 2 inches in diameter. The case covers two sections taped together. The upper section is a protective cap removed prior to loading; the lower section is loaded into the firing drums of the simulator and contains the pyrotechnic charge and igniter. This simulator is yellow or olive with white markings.

Figure 2–12 Chemical bombs.

■ CHEMICAL ORDNANCE

Chemical ordnance can be as simple as a riot grenade (tear gas) or smoke bomb or as lethal as sarin nerve agent. U.S. chemical weapons are typically gray with colored bands. These colored bands indicate what type of filler is inside the ordnance. The color code has changed over time, and other countries have different color

Figure 2–13 Suitcase bomb.

codes. (See Chapter 4 on Weapons of Mass Destruction and the chart in the color insert for more detail.) Chemical weapons use a small amount of high explosives to burst the container and change the liquid to gas or vapor for greater effect.

■ NUCLEAR WEAPONS

There is nothing exotic looking about nuclear weapons or "special weapons", as the military calls them. Tactical weapons such as artillery shell, air-to-air missiles, short

Figure 2–14 Backpack bomb.

range missiles, and some air-dropped weapons are being phased out or are already gone from service.

Suitcase bombs hyped by the media were really not suitcase sized at all. They were called Atomic Demolition Munitions (ADM) and came in two sizes. The Medium Atomic Demolition Munition (MADM) had the W45 warhead and weighed about 400 lbs, while the Special Atomic Demolition Munition (SADM) had the W54 warhead and weighed 58 lbs. The SADM was also nicknamed the backpack bomb. These were produced in the 1960s and phased out in the 1980s. There are reports in the media that several of the Russian versions are missing. Chapter 4 will discuss more about nuclear weapons.

■ CONCLUSION

Military ordnance should always be left alone by untrained personnel. Those who find explosives can note their colors and markings, but only if it can be done without disturbing the ordnance and at a safe distance. When ordnance is found, the local civilian bomb-disposal team or military bomb-disposal authorities should be contacted. The devices should be treated like any hazardous material incident.

▪ 3 ▪

BOMB THREATS AND BOMB SEARCH PROCEDURES

Many of today's bombing incidents involve improvised explosive devices (IEDs) or homemade bombs. First responders should know about the triggering mechanisms of such bombs as well as about bomb threats and bomb searches.

▨ TRIGGERING MECHANISMS OF IMPROVISED EXPLOSIVE DEVICES

Triggering mechanisms of IEDs can be located inside or outside small-sized or large-sized bombs. Bombs can look like explosive devices or can be disguised as such objects as books, briefcases, letters, or packages.

Figure 3–1 Possible bombs.

Because there are a variety of triggering mechanisms, it is important to leave untouched any suspicious looking object, isolate the area, and notify the appropriate bomb-disposal authorities. Switches come in all shapes and sizes.

A **trembler switch** is a device that senses vibrations around bombs. It may be made to trip after a certain number of vibrations.

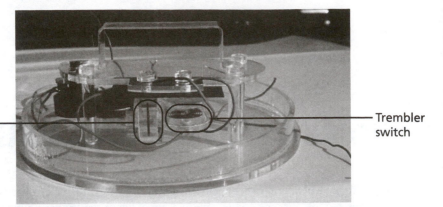

Trembler switch   Trembler switch

Figure 3–2 Trembler switch on Russian land mine; one vertical, one horizontal.

An **antiprobe switch** prevents any metallic object, like a knife, from cutting into a bomb. Surrounding the bomb are two conductive layers of aluminum foil, with a nonconductive layer of such material as paper in between. A metallic object that cuts through the two conductive layers completes the circuit, detonating the bomb.

A **barometric (baro) switch** completes the circuit when a bomb is subject to a certain atmospheric pressure. The most common baro switch is a balloon encased in a container. As the balloon expands, it pushes a conductive plate into two electrical leads and completes the circuit.

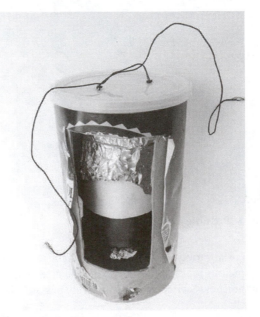

Figure 3–3 Baro switch.

A **thermostatic switch** operates at a given temperature and can be as simple as a wall thermostat.

Figure 3–4 Thermostatic switch.

A **collapsing relay circuit** detonates a bomb if any wire in the circuit is cut. A good knowledge of electronics is required to make this triggering device.

Figure 3–5 Stud finder switch.

A **magnetic switch** operates if any metallic item is passed over or near the bomb. A simple stud finder will work.

A **photoelectric switch** operates if a light is shone into a package or if a package is opened, exposing the inside to light. Photoelectric switches can be bought in almost any electronics store.

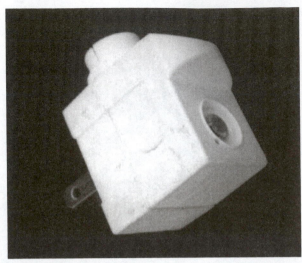

Figure 3–6 Photoelectric switch.

An antiopening switch will detonate a bomb if the package is opened or if a drawer or door is pulled open. A simple antiopening switch can be made with a clothespin and two thumbtacks wired to its open jaw. The jaws are held open by a wooden peg. The peg is either glued to the top of a lid or tied by a string to a door or drawer. As the peg is pulled out, the two jaws snap together and the circuit is complete, detonating the bomb.

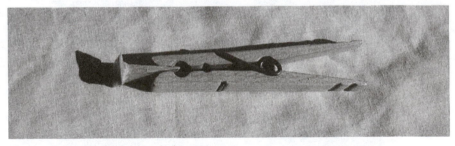

Figure 3–7 Clothespin switch.

A timing switch can be nothing more than a clock that, at a set time, activates the rest of a circuit. This method allows a bomber time to place a bomb without setting it off prematurely. Once the bomb is hidden, the bomber sets the timer to allow enough time to leave the area. When the timer runs down, the bomb is armed. A second timing switch is used to activate the bomb.

An antidisturbance switch trips if the device is moved or tilted. A mercury switch is often used.

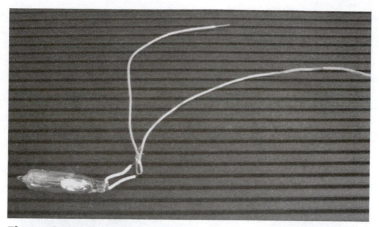

Figure 3–8 Mercury switch.

A radio-controlled switch is activated by a radio transmitter. These can be purchased at most hobby shops, which sell them for radio-controlled model airplanes and cars.

Figure 3–9 Radio-controlled switch.

Sound-activated switches are found in toys and other devices that turn on and off with sound.

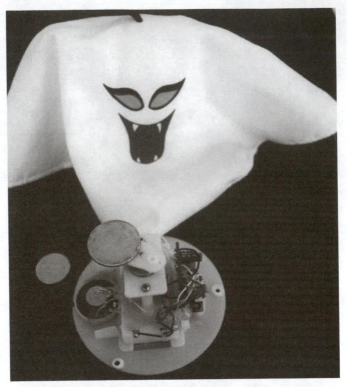

Figure 3–10 Sound-activated switch.

■ BOOBY TRAPS AND ANTIPERSONNEL DEVICES

Booby traps and antipersonnel devices have been used against police and firefighters. A common place to find booby traps are in illegal drug labs. Any bomb disguised as something other than a bomb is a booby trap.

The firefly, which is technically not an explosive, consists of a gelatin capsule filled with calcium carbide and sodium that is dropped into a gas tank of a police car or fire truck. There is usually a little water in the bottom of most gasoline tanks due to condensation. When the water in the gas tank dissolves the capsule, the calcium carbide mixes with the water to form acetylene gas. The sodium reacts with the water to burst into flame and ignite the acetylene gas. The resulting explosion ruptures the gasoline tank.

A shotgun shell with a BB taped over its primer and a fin or streamer on the top is sometimes dropped from rooftops onto firefighters below. These shotgun shells were used during the race riots in the 1960s in cities such as Chicago.

Fire extinguishers can also become explosive devices. For example, during an arson fire in a San Antonio pizza parlor, a fire extinguisher filled with gasoline was discovered. Fortunately, there was no pressure in the extinguisher because the powder had been removed.

Figure 1   Blasting Caps

Figure 5   Ammonium Nitrate

Figure 2   Blasting Caps

Figure 6   Blasting Agent

Figure 3   Safety Fuse

Figure 7   Kinepax

Figure 4   Detonation Cord

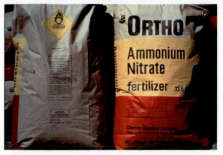

Figure 8   Booster

Figure 9   Boosters

Figure 14   Types of Dynamite

Figure 10   Dynamite

Figure 11   Dynamite

Figure 15   Slurry

Figure 12   Dynamite

Figure 16   Dynamite and Slurry

Figure 13   Dynamite

Figure 17   Smokeless and
                    Black Powder

Figure 18   Fuzes

Figure 22   Tail Fuzes

Figure 19   Nose Fuzes

Figure 23   Tail Fuze

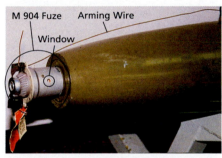

Figure 20   M 904 Fuze

Figure 21   Nose Fuzes

Figure 24   VT Fuze

Figure 25   VT Fuze

Figure 28   Projectile Fuzes

Figure 26   VT Fuze

Figure 29   Hand Grenades

Figure 30   Hand Grenades

Figure 27   "Rosette" Fuze

Figure 31   Hand Grenades

Figure 32   Rifle Grenades

Figure 36   Projectiles

Figure 33   Rifle Grenades

Figure 34   Rocket Propelled Grenades

Figure 35   Projectiles

Fuze

Rotating
Bands

Figure 37   155 mm Projectile

Figure 38   Projectile

Figure 39   Sabot

Figure 40   Flechettes

Figure 41   Mortars

Figure 42   Mortars

Figure 43   Mortar, Japanese

Figure 44   Russian RPG

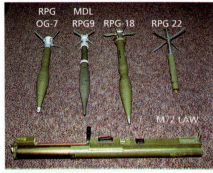

Figure 45   RPG's

Figure 46   Rocket and Rocket Pod

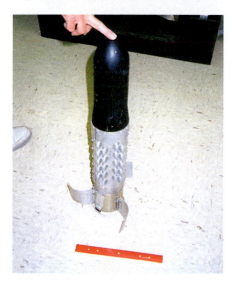

Figure 47   Rocket Pod and Rockets

Figure 48   Dragon Missile

Figure 49   Dragon Missile

Figure 50   Stinger Missile

Figure 51   Stinger Missile

Figure 52   TOW Missile Pod

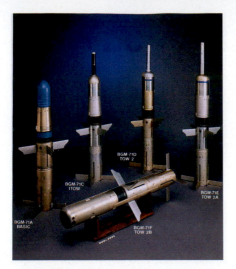

Figure 53   TOW Missiles

Figure 54   Missiles

Figure 55   Missiles

Figure 56   Land Mines

Figure 57   Claymore Mine

Figure 58   Gator Mine

Figure 59   M 7A2 Mine

Figure 60   Chinese Type 72 Land Mine

Figure 61   PMN-2 Mine

Figure 65   Bounding-Type MInes

Figure 62   PMN Mine

Figure 66   AT Mines

Figure 63   VS-50 Mine

Figure 67   Submunitions

Figure 64   Mines

Figure 68   BLU-54/B

Figure 69   BLU-3 Submunition

Figure 70   Russian Submunition

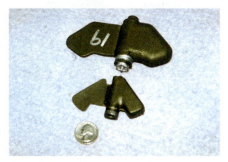

Figure 71   PFM-1 (top), Dragon Tooth Submunition

Figure 72   AT BLUs (U.S.) Submumition

Figure 73   AT BLU Submunition

Figure 74   BLUs and CBU

Figure 75   CBU-59/B with BLU-77/B

Figure 76   CAD Item

Figure 77   Cartridge

Figure 78   Cartridge

Figure 79   Cartridge

Figure 80   Air-Dropped Bombs

Figure 81   MK-82 Bomb

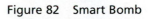

Figure 82   Smart Bomb

Figure 83   Practice Bombs

Figure 84   Frag Bomb

Figure 85   Chemical Bomb

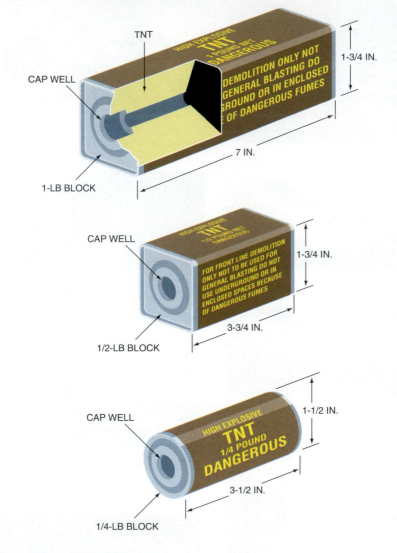

Figure 86   Bulk Explosives

Figure 87   Bulk Explosives
              C-3

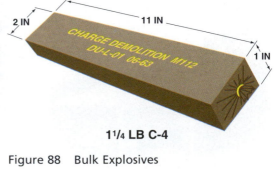

1¼ **LB C-4**

Figure 88   Bulk Explosives

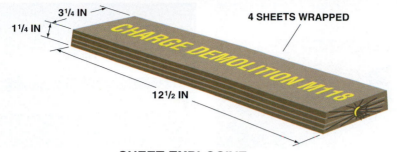

3¹/₄ IN

1¹/₄ IN

4 SHEETS WRAPPED

CHARGE DEMOLITION M118

12¹/₂ IN

SHEET EXPLOSIVE

Figure 89   Sheet Explosive

Figure 90   Fuse Lighter

Figure 91   Incendiary Bomb

Figure 92   WWII Bombs

Figure 93   Napalm Bomb

Figure 94   Napalm Igniter

Figure 95   Fuel Air Explosive (FAE)

Figure 96   Flare Dispenser

Figure 97   Flare

Figure 100   Flare

Figure 98   Chaff

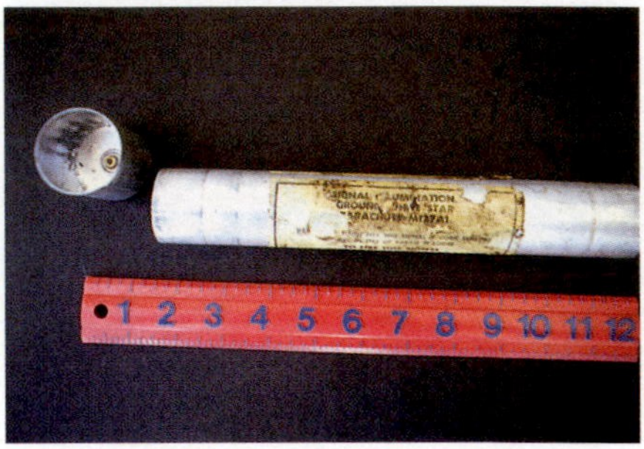

Figure 101   Flare

Figure 99   Flare

Figure 102   Simulators

CHEMICAL MUNITIONS MARKINGS
FIVE ELEMENT MARKING SYSTEM (709A)

BACKGROUND (CML MUNITION)	NO. OF BANDS (DURATION OF EFFECTIVENESS)	COLOR OF MARKINGS[1] (PRIMARY USE)	CHEMICAL AGENT SYMBOL (EXACT FILLING)	DESCRIPTIVE WORDS (GENERAL NATURE OF AGENT ON RELEASE)
	(one green band)	TOXIC CHEMICAL AGENTS (CASUALTY AGENTS)	GB, CG, CK	GAS
	(two green bands)	TOXIC CHEMICAL AGENTS (CASUALTY AGENTS)	VX, HD, H, HT	GAS
	(one red band)	IRRITANT AGENTS (RIOT CONTROL AGENTS)	CN, CS CN1, CS2	GAS
	(one purple band)	INCENDIARIES	TH, NP, PTI, PTV	INCENDIARY
	(one yellow band)	SMOKES (SCREENING AND SIGNALING)	HC, WP, PWP,	SMOKE

[1] Markings include bands and lettering

STANDARD COLOR CODING SYSTEM (709B)

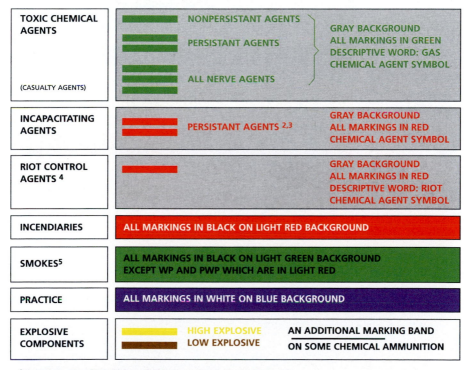

TOXIC CHEMICAL AGENTS (CASUALTY AGENTS)	NONPERSISTANT AGENTS PERSISTANT AGENTS ALL NERVE AGENTS	GRAY BACKGROUND ALL MARKINGS IN GREEN DESCRIPTIVE WORD: GAS CHEMICAL AGENT SYMBOL
INCAPACITATING AGENTS	PERSISTANT AGENTS [2,3]	GRAY BACKGROUND ALL MARKINGS IN RED CHEMICAL AGENT SYMBOL
RIOT CONTROL AGENTS [4]		GRAY BACKGROUND ALL MARKINGS IN RED DESCRIPTIVE WORD: RIOT CHEMICAL AGENT SYMBOL
INCENDIARIES	ALL MARKINGS IN BLACK ON LIGHT RED BACKGROUND	
SMOKES[5]	ALL MARKINGS IN BLACK ON LIGHT GREEN BACKGROUND EXCEPT WP AND PWP WHICH ARE IN LIGHT RED	
PRACTICE	ALL MARKINGS IN WHITE ON BLUE BACKGROUND	
EXPLOSIVE COMPONENTS	HIGH EXPLOSIVE LOW EXPLOSIVE	AN ADDITIONAL MARKING BAND ON SOME CHEMICAL AMMUNITION

[2] Currently munitions filled with incapacitating agents are marked as persistant agents.
[3] No descriptive word is on incapacitating agent filled munitions.
[4] ENSURE munitions containing CS2 fill will be marked with two red bands to denote a persistant agent.
[5] M18 colored smoke hand grenades have an alternate green (OD) base color with lettering and a 1-inch band of light green to show primary use.

REVISED COLOR CODING SYSTEM (709C)

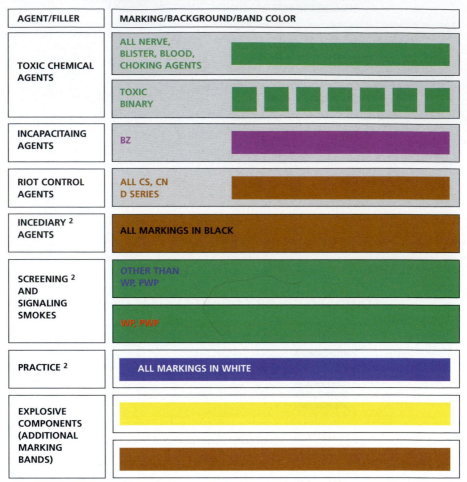

AGENT/FILLER	MARKING/BACKGROUND/BAND COLOR
TOXIC CHEMICAL AGENTS	ALL NERVE, BLISTER, BLOOD, CHOKING AGENTS
	TOXIC BINARY
INCAPACITAING AGENTS	BZ
RIOT CONTROL AGENTS	ALL CS, CN D SERIES
INCEDIARY [2] AGENTS	ALL MARKINGS IN BLACK
SCREENING [2] AND SIGNALING SMOKES	OTHER THAN WP, PWP
	WP, PWP
PRACTICE [2]	ALL MARKINGS IN WHITE
EXPLOSIVE COMPONENTS (ADDITIONAL MARKING BANDS)	

[1] Markings include name of chemical agent symbol.

[2] No band included.

Figure 3–11 Fire extinguisher.

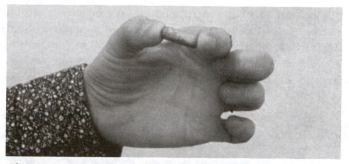

Figure 3–12 Firefly gelatine capsules.

■ BOMB THREATS

The increasing number of bombings involving IEDs requires the public to know more about bomb threats and bomb searches. Today, any individual or organization may be involved in a bomb threat or bomb search. Those in authority must be especially prepared for bomb threats. They should understand the motives of bombers and how they tend to operate. Bomb threats are usually communicated by telephone or mail. Bombs are rarely delivered in person, although this does happen.

Motives of Bombers

A bomber's motive for his or her actions can be important for authorities to know. A bomber's motive can provide clues about his or her intended target, the type of bomb being used, and the location of the bomb.

Revenge is a common reason for bombings. Feelings of revenge can come from a fired or a disgruntled employee as well as from a person involved in a domestic dispute. A bomb threat based on revenge is usually aimed at a specific person or object.

Jilted lover hands grenade to woman

A man pulled the pin of a grenade and handed it to his ex-girlfriend, then watched as she ran away from nearby houses in Battle Creek, Mich., and held onto the safety lever until police arrived.

No one was injured when Lisa Wood fled from the steps of her home to an open area in her neighborhood — followed by her former boyfriend in his car — then knelt down and waited for police, police said.

Said Dave Durham, a neighbor who witnessed the incident: "I yelled to her, 'Lisa! Don't let go! Hold it baby, hold it! Just sit down and wait till the police get here!' And she was squeezing it. That little girl had a death hold on that grenade."

A 30-year-old man was being held at the city jail. He is expected to be arraigned Monday on assault charges, Sgt. Beverly Palumbo said. Police did not identify him pending the formal filing of charges.

"She was afraid but she remained rational," Detective Tim

> **"** I yelled to her, 'Lisa! Don't let go! Hold it baby, hold it!' **"**
>
> — Dave Durham, neighbor

Hurtt said Saturday. "Officers were able to put something in the place of the pin."

Police said they confiscated a second grenade at the suspect's home. A state police bomb squad was dissecting the grenades this weekend to determine whether they were live, Palumbo said.

Wood and the suspect dated a few weeks ago, but she broke up with him, said Wood's friend Frances Hall.

Police arrived within minutes after the suspect handed Wood the grenade on the steps of her home, authorities said.

Figure 3–13 Newspaper article on revenge.

Politically motivated and terrorist bombings are increasing worldwide. With all the political unrest today, this situation may not improve. Some political groups use terrorist techniques to spread their political message. Although these groups claim to be the true representatives of the people, they often strike unarmed civilians and nonmilitary targets.

18 car bombs seized at shopping center

Two ex-Green Berets nabbed in San Antonio

By JEROME P. CURRY

Federal agents posing as Mexican terrorists snared two former Green Berets and confiscated 18 car bombs in a crackdown on international terrorism.

Also facing charges is the wife of one of the former special forces men as agents made nearly simultaneous arrests at a San Antonio shopping center parking lot and a house at

Round Rock at 11:30 a.m. Saturday.

One official said the arrests were part of a new emphasis on cracking down on international terrorists from both the right and left trafficking in explosives in the southwestern United States.

Law enforcement agents in at least four nations, the U.S., Mexico, Guatemala and Honduras, have noted this increase.

The 18 bombs designed to blow apart automobiles were confiscated in the San Antonio arrest at Interstate 410 and Blanco Road.

No resistance or trouble was of-

fered at either of the arrest sites, said Robert Rowe, special agent in charge of the San Antonio office of Alchohol, Tobacco and Firearms.

Rowe said the Saturday raids climaxed 30 days of negotiations between ATF agents disguised as Mexican terrorists and the three people charged with bomb-making.

James A. Paxton, 43, a former army special forces major; his wife, Frances, commonly called Frankie, 38, and David Thomas Nicewander, 33, identified by Rowe as a former special forces operative, were ar-

rested and face federal charges, authorities said.

Paxton was identified as Nicewander's former commander.

Federal officials said Paxton faces charges on 34 counts of sale, possession and manufacture of explosive devices. He is in Bexar County Jail.

Officials said Mrs. Paxton faces charges on one count each of possession and sale of explosive devices. Nicewander faces charges on one state transportation of explosive

See BOMB, Page 7-A

Figure 3–14 Newspaper article on terrorists.

Insurance was once a popular reason for bombing, but it has since decreased as a motivator. For a while, it was in vogue to insure a family member for a large amount of money and then to send the person somewhere by airplane with a bomb in his or her suitcase. The use of explosives to destroy persons or property to collect insurance is rare today, however.

Extortion, the act of gaining money or favors through the threat of harm, is the motivation of some bombers. These bombers threaten to destroy a person or property but offer to reveal the location of the bomb if they receive money. They often tell bomb threat recipients the location of the bomb, whether or not they receive payment for the information, so that recipients take the threats seriously. Then the bombers often claim that they have other bombs planted and demand more money for information about their location.

Coercion, or forcing others to do something against their will, is another motive for bombings. In addition, organized crime uses bomb threats or bombings to force businesses to participate in protection rackets or to eliminate competition. These types of bombings, however, are designed to scare others and usually do not cause death or damage. Those who make threats want to avoid destroying their source of income.

Mental Disorders

Some psychopaths who suffer a character disorder distinguished by amorality or antisocial behavior are also motivated to use bombs. Their reasons are varied, and only trained professionals can explain this behavior. Sometimes a need for power is the reason. Seeing their actions reported on television and radio or in the newspaper can encourage psychopaths and others to commit more crimes. For this reason, media coverage of bomb threats or bombings should be restricted. The public's need to know must be weighed against law enforcement's desire to prevent future bombings.

First responders are often depicted as rescuers, yet, they are as vulnerable to bomb threats as those they protect. Because such personnel represent the government to bombers, they can also be targets.

First responders should be careful and must not allow the acts of a few individuals to destroy their sense of service to the community.

Phone Threats

If a bomb threat is communicated by phone, the recipient should keep the caller talking as long as possible. The recipient should ask the caller to repeat the message, and should try to maintain his or her composure. Many times, the recipient of a bomb threat loses the wording of a caller's initial message due to the shock of the call.

A recipient of a phone threat should ask when the bomb is going to explode. Time is critical. A bomb threat recipient should also ask where the bomb is located and what it looks like. A bomber is often proud of his or her work and may brag about the bomb. The recipient of such a threat should also inquire about the reason for the bomb. The response may help establish the bomber's motive and target.

While on the phone, a recipient should try to identify background noise. The sound of music, traffic, aircraft, voices, or other noises may provide clues to the location of the call. He should pay close attention to the voice of the caller for information concerning the caller's age and gender. He should note if the caller's voice sounds calm or excited, if it is recognizable, or possesses any distinguishing speech characteristics. The time a caller hangs up should also be recorded by the recipient and reported to the police, along with his or her name, address, and phone number so that the authorities can contact the recipient of the call later, if necessary.

ATF BOMB THREAT CHECKLIST

Exact time of call _____

Exact words of caller _____
_____
_____

QUESTIONS TO ASK

1. When is the bomb going to explode?_____
2. Where is the bomb?_____
3. What does it look like?_____
4. What kind of bomb is it?_____
5. What will cause it to explode?_____
6. Did you place the bomb? _____
7. Why? _____
8. Where are you calling from? _____
9. What is your address?_____
10. What is your name? _____

CALLER'S VOICE (circle)

Calm	Stressed	Deep	Sincere	Angry	Slurred	Excited
Stutter	Disguised	Accent	Crying	Lisp	Broken	Normal
Giggling	Slow	Nasal	Loud	Squeaky	Rapid	

If voice is familiar, whom did it sound like?_____

Were there any background noises? _____
Remarks: _____
_____

Person receiving call: _____
Telephone number call received at: _____
Date: _____
Report call immediatly to: _____
(Refer to bomb incident plan)

Figure 3–15 Bomb threat checklist.

PACKAGE

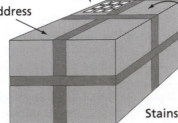

Stamps, rather than metered
Excessive postage, foreign postage
or cancellation stamp

No return address

Wrong title with name
or misspelled name

Stains on package
or strange odors

Excessive tape, string
or packaging

Figure 3–16 Package bomb.

Figure 3–17 Letter bomb.

Mail Threats

Bombs sent by letter or package through the mail or through a delivery service may involve any of the previously mentioned motives. Mail bombs sometimes have false return addresses or no return address.

Those who believe that any letter or package looks suspicious should avoid handling it. They should clear the area and phone police authorities.

■ BOMB SEARCHES

Planning

Organizations subject to a bomb scare should have a plan of action. Lines of authority must be established to determine who does what. With such a plan, confusion and panic can be prevented. Panic is often caused when people do not know what to do and what is happening. Calm leadership is the best way to prevent panic.

Managers of organizations should first contact their local police and fire department personnel for help in developing bomb search plans. Such managers should determine who has a bomb-disposal unit and whether or not those in their own organization will be involved in the search process.

Many fire departments do not become involved in the process. Instead, they are on standby in case a bomb does go off, ready to extinguish any resulting fires and provide rescue services. If emergency crews are disabled in an initial bomb blast, they are of little use. In addition, emergency crews may not be familiar with the specific area and would not know what belongs in the building. Only those who live or work in the involved area have this knowledge. Bomb search teams consisting of members of the company or residents of the area should be formed. Such teams should conduct searches of areas in buildings where explosives might be placed.

A traditional fire escape plan is needed in case of evacuation. Once the plan is developed, it must have the approval and support of management, particularly when it comes to drills and training.

In some bomb search classes, a fake bomb is planted somewhere in the company holding the class. The class then breaks into teams to conduct a practice search. Most effective in finding bombs are the personnel who work in that area.

Figure 3–18 Search teams must have keys to every room and closet.

Search teams should be thoroughly trained and be familiar with their search areas. Assignment to a search team should be voluntary. These activities, however, involve the loss of work time and require the cost of equipping the search teams. In addition, managers should provide some compensation for such service, either in time off or extra pay.

Usually, only after an organization has received a bomb threat do its leaders become concerned. At this point managers seek employee training. Unfortunately, as time passes without further threats, the training and the subject seem to slip from the minds of some managers.

Equipment

A minimum amount of equipment is necessary to search for bombs. The following items may be helpful.

A flashlight is a requirement because many areas to be searched will be dark or searches will be done at night. Extra batteries should be available. Screwdrivers are needed to gain access to ventilation panels and other equipment.

Search teams must have keys to every room and closet in their area of responsibility.

A fishhook tied to a fishing line can be used to open doors and drawers remotely.

Figure 3–19 Grappling hook opening a drawer remotely.

Several hooks can be soldered together to form a grappling hook. Picture frame hooks or C hooks can be used with the fishing line to act as a fulcrum. Care must be used because the hooks may permanently damage the surface to which they are attached.

Figure 3–20 Grappling hook and fish line.

Tape may be used for several purposes. Searchers can tape over windows that cannot be opened. Windows near a suspected bomb should be opened or taped to prevent the glass from fragmenting from a blast. Tape can also be used to mark a room that has been searched.

A stethoscope can be used to listen to drawers for the ticking of a bomb's clockwork before the drawer is opened or a door or a panel is removed. Many bomb triggering mechanisms do not make any sounds, however.

General Search Procedures

Search teams should work in pairs. They should be so familiar with the area to be searched that they can identify any object that does not belong there. They should divide the building to be searched into zones. The teams should then be assigned to certain zones. The outside of a building must also be searched and should be designated a zone in itself.

A command post should be established in a secure area with telephones for communications. *Two-way radio transmissions cannot be used* because the electromagnetic radiation (EMR) from them may detonate the bomb.

Fire and EMS personnel should stand by at a safe distance to respond quickly in case of a bomb's detonation.

Once a command post has been established and search teams have been given their assignments, the search should begin. Where to start is usually a matter of judgment. The evacuation path and the evacuation area, however, should be searched first to provide a safe path from the building. In searching the outside of a building, a team should examine all drainage pipes, mailboxes, shrubs, trash cans, and vehicles. Utility connections such as gas and electricity should also be searched. Electrical generators, transformers, and boiler rooms are prime targets. Bombers think about hiding bombs in places that will do the most damage.

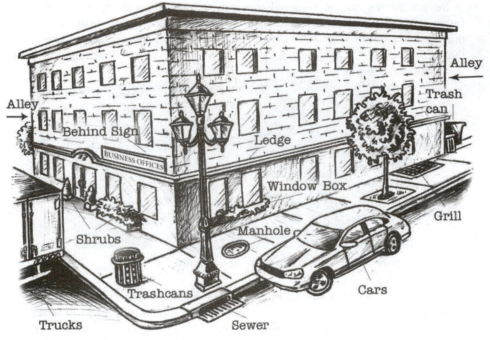

Figure 3–20 Outside search.

In searching vehicles, teams should look under fenders and under vehicles. Mirrors can assist them in looking for trip wires to a hood, door, or trunk. Latches on car parts can be opened remotely by taping open the latches of such parts and using rope to pull them open. This method can be extremely hazardous, however, because any movement can set off a bomb. In addition, a bomb may be detonated remotely by a bomber who is watching from a distance.

Once teams have searched the outside of buildings and the evacuation area, they should search the command post area along with hallways, lobbies, rest rooms, and other places accessible to the public. In a rest room, a bomber may place bombs in trash containers or used sanitary napkin containers, knowing the reluctance of the search team to search such areas. A broomstick can be used to search through the materials in such containers without using one's hands. Gloves should also be used.

Removable ceiling panels are ideal areas in which to place bombs. Searchers should look for ceiling tiles that are out of place as well as for dust on the floor or furniture. Dust may have resulted from the bomber's removing the ceiling tiles, planting a bomb, and then replacing the tiles.

Search team members should also look for recently removed screws. If the screws are painted over, they have not been removed. If the paint on the heads of the screws appears to be scraped or broken, one can suspect a bomb inside the panel.

Conducting a Room Search

With teams, each room can be divided into two zones. The bottom zone consists of everything from the floor level to eye level. One search team should look under car-

Figure 3–21 Celing tile out of place.

Figure 3–22 Dust on desk.

Figure 3–23 Painted-over screw.

pets, furniture, and lamps and in trash containers. The top zone starts at eye level and extends to the ceiling. Search team members should look behind curtains and at window ledges. They should inspect light globes and remove suspected ceiling tiles to check the ceiling space. They should also inspect air vents for signs of tampering.

After a team has completed its search in its zone, the team should switch zones with the other team and repeat the process. After both zones have been searched by all members of the search team, they should use tape or chalk to indicate that the room has been searched.

If team members find suspicious-looking objects, they should avoid moving or touching them. Instead, they should notify the command post, either by telephone or a means *other than two-way radio*. They should provide the location and a description of the objects.

Figure 3–24 Room divided into two zones: zone one from floor to eye level, and zone two from eye level to ceiling.

Search team members should then open doors and windows around the object to dissipate the blast wave if a bomb does go off. If windows cannot be opened, the face of the window should be taped to prevent the glass from fragmenting into tiny slivers.

The team should then evacuate the immediate area above, below, and adjacent to the bomb site. The size of the suspected bomb, the construction of the room, and other factors will determine the size of the danger zone.

A team should continue to search even after a suspected bomb is found. There may be secondary or other bombs, or the object found may be a decoy. Only after all areas have been searched is the work of a search team completed.

Evacuation

The decision to evacuate is a difficult decision to make. Only the authority responsible for the people under its care, such as school principals and corporate managers, should make this decision, not the emergency personnel. One major problem occurs when a bomb threat involves an office building housing many different businesses. Establishing who is the authority in charge is sometimes difficult. In such a situation, public safety personnel may have to assume the responsibility for making the decision to evacuate.

Such a decision depends on many factors. Among the most important is the information from the bomber's phone call and the authenticity of the call. If a suspected bomb is found, the size of the bomb becomes a factor.

In addition, one must consider the type of building and the quality of its construction. In a well-constructed, fire-resistive building, it takes an enormous number of explosives to cause a structure to collapse. A large amount of explosives would be difficult to hide and would be readily seen if located inside the building. A vehicle bomb such as was used in the World Trade Center and the Oklahoma City bombings is different. The decision to evacuate a building should be considered with care. The occupants of a well-constructed building are often safer remaining in a room of the building than exiting hallways or standing around outside the building. A bomber may have planted a bomb in the hallway or at the evacuation site.

Predetermined evacuation routes may have to be changed, depending on the location of the suspected bomb. The evacuation area should be safe, secure, and out of the weather for the comfort of the people. Personnel should not be allowed to linger in front of or around the building and should not be let back into the building until permission has been given by the proper authorities. In far too many bomb threat situations people are allowed to stand outside in front of building, with glass windows that could explode, or next to cars in parking lots which could contain a bomb either as the primary or secondary explosive. Figure 3–26 shows what is left of three cars in a hotel parking lot. The explosions blew out all the front windows. Fortunately, only one person was slightly injured.

The decision to turn off gas and electrical power during a search must be made with the requirements of the search teams in mind.

Figure 3–25 Car bomb.

■ CONCLUSION

This chapter has been only a brief discussion and guide on bomb threats and bomb searches. For more information, there are several good books on this subject and the local ATF or bomb squad will have more information. First responders need to know about triggering mechanisms, bomb threats, and bomb searches in order to better assist and protect the public.

■ 4 ■

WEAPONS OF
MASS DESTRUCTION (WMD)

First responders are faced with many and varied hazards every time they respond to an emergency. They can face fire, explosion, building collapse, auto accidents, medical emergencies, and law enforcement incidents. In addition to this laundry list of hazards, we must add another: weapons of mass destruction, or WMD. This new hazard comes from both international and domestic terrorists; these terrorists can be from rogue nations, rebel groups, or lone individuals. Today's technology has increased the destructive power that these groups or individuals can possess. In 1995, I participated in a New York City exercise, involving Sarin nerve agent in a subway. It was estimated that the first one hundred responders would have been killed by rushing in without knowing what they faced. It also showed the need for further planning and exercises involving multiple agencies.

Figure 4–1 New York City WMD exercise.

Although the new buzz word for WMD is CBRNE (chemical, biological, radio-logical, nuclear, and explosive) for our purpose, I will break WMD into just three categories to keep it simple. These are nuclear, biological, and chemical (NBC).

■ NUCLEAR WEAPONS

Nuclear weapons fall into two categories: nuclear weapons and radiological dispersal devices (RDD).

WEAPONS

Nuclear weapons are probably the least likely to be used. Terrorist would have to steal one from a country that possesses them. It is unlikely that a weapon could be stolen, except from a previous Soviet bloc nation since the security there is lacking.

There are two types of nuclear weapons, the A-bomb and the H-bomb. An A-bomb uses a fissile material in which the atoms are split, causing a chain reaction; and an H-bomb, uses material in which the atoms are fused together. It takes an A-bomb to trigger an H-bomb so really an H-bomb is two bombs in one. A terrorist would more than likely use an A-bomb since it is easier to make using the gun type.

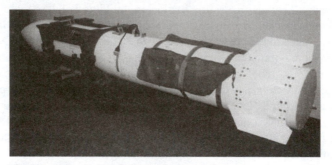

Figure 4–2 B-83 Thermonuclear weapon.

A terrorist weapon would probably be a gun-type weapon rather than an implosion-type weapon. In a gun type, there is a doughnut of nuclear material, which has a slug of nuclear material shot into the hole by an explosive charge. This causes the material to go supercritical and a nuclear explosion results. There is no reaction as long as the doughnut and slug are separated.

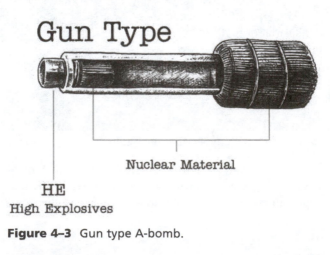

Figure 4–3 Gun type A-bomb.

In an implosion-type weapon, the nuclear material is in a sphere shape surrounded by a sphere of high explosives (HE). The HE is detonated at numerous points around the sphere, and the implosion wave from the HE compresses the nuclear material into a small volume causing it to go supercritical. The result is an atomic explosion. This method is extremely difficult to achieve, and it is unlikely that a terrorist or rogue nation could pull it off. A gun type is easier to make.

Implosion type

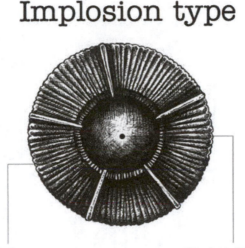

HE
High Explosives

Nuclear Material

Figure 4–4 Implosion type A-bomb.

RADIOLOGICAL DISPERSAL DEVICE

A more likely situation would be for terrorists to make their own weapon using nuclear material, stolen or bought. A radiological dispersal device (RDD), commonly called a dirty bomb, is a conventional explosive used to spread a radioactive material around, contaminating the area. This radioactive material could even be material used in industrial or in a medical lab. Over 19,000 sources of this type of radiation exist in the U.S. alone. It would not be hard to obtain. The area affected by RDD would be small compared to other devices and limited by the amount of explosive and nuclear material.

Radiation is tasteless, odorless, and invisible. It can only be detected by special radiaic instruments. There are three types of radiation from a radioactive material.

The first is the **alpha particle**, which is a Helium nucleus with a positive charge on it. It is a heavy particle. This means it does not travel very far (an inch or two) before it meets a stray electron and becomes Helium gas and becomes quite harmless. Alpha radiation is an internal hazard only, which means it must be ingested or inhaled to be hazard. It will not penetrate clothing or skin.

Beta radiation is a high-energy electron. It cannot penetrate heavy clothing, so it, too, is basically an internal hazard.

Gamma radiation is the most dangerous of all radiation. It is an electromagnetic wave similar to x-rays, but with much more energy and penetrating power. It takes several inches of lead to stop a gamma ray.

PROTECTION FROM RADIATION

Shielding: Normal clothing can stop both alpha and beta radiation. Unless inhaled or eaten, normal clothing with some type of filter mask will be proper protection.

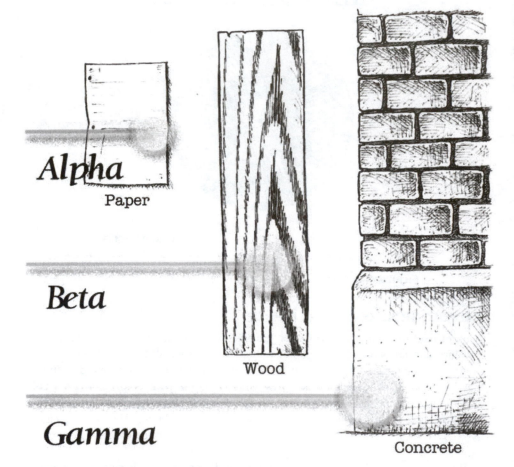

Alpha
Paper

Beta
Wood

Gamma
Concrete

Figure 4–5 Radiation Penetration.

Time: The protection with time falls into two categories. The first is called the half-life of a material. This is the time it takes for one-half of the material to decay. It may be in seconds or millions of years, depending upon the material. For example, if 10 lbs of element X had a half-life of 1 hour, at the end of 1 hour, there would be only 5 lbs left of X. The other 5 lbs would be of another element.

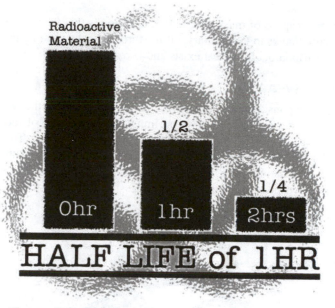

Figure 4–6 Half-life.

The second category is exposure time. This is the time spent in a radioactive area times the exposure rate. For an example, if a rate of radioactive is 25 mrem/hr and you spend one hour in the hot zone, your exposure will be 25 mrems.

Distance: Distance is probably the best protection a first responder can have. The radiation falls off the square of the distance. This means the radiation at 5 feet away is 1/25 or 5 squared of that a source. If a source is 100 r/hr, at 5 feet it is only 4 r/hr.

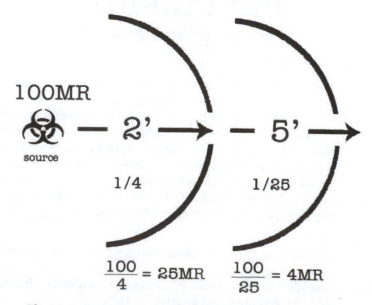

Figure 4–7 Radiation vs. Distance.

Of the three weapons of mass destruction, radiation is the one we have the most experience with, thanks to the military. Of the three, it is the least likely to be used as a terrorist, weapon but the threat exists and is real.

■ BIOLOGIAL WEAPONS

Biological weapons (sometimes called germ warfare) are the worst of the weapons of mass destruction. There is no immediate indication of their use. The incubation period can run 2-6 days before symptoms begin to show. The symptoms may be similar to those of the flu. Biological agents are mostly an inhalation and ingestion hazard, although some can attack an opening in the skin and through the eyes. Biological weapons are nonvolatile, tasteless, and invisible to the eye.

Figure 4–8 Biohazard symbol.

BACTERIA

Anthrax is found worldwide and is primarily a disease of livestock such as cattle, horses, sheep, and goats. The spores are long lasting and can be dangerous for years in the ground and water under the right conditions. It can even resist the ultraviolet light in sunlight for several days. The most common infection is a result of a skin infection or lesion from contact from people who have had contact with infected animals. Inhalation of the spores is the most dangerous method of contraction.

The World Health Organization computer model says if 110 lbs of Anthrax was sprayed along a 1.5 mile track, upwind of 500,000 people, 24,000 of them would die. The mortality from inhalation of anthrax is 80-90 percent.

Early symptoms:
- 1-2 days—Symptoms are like those of the flu with fever, chills, headache, and sometimes chest pains.
- 3-5 days—Symptoms include cyanosis (bluing of skin), respiratory distress, rapid heart rate, and low blood pressure.

Cholera

Cholera is not easily spread from human to human; it is usually spread in contaminated drinking water or sabotaged food or water supply. Chlorine will kill it, as well as boiling the water or temperature of 118 degrees or better. The incubation period is 12-72 hours and the symptoms are vomiting, headache, cramping, and severe diarrhea. There is a 50 percent mortality rate if untreated.

Plague (Black Death)

There are three types of plague: bubonic, septicemic, and pneumonic. It is transmitted from rats to man by fleas. The incubation period is 2-10 days for bubonic and 2-3 days for pneumonic. Its symptoms are similar to the flu. At near freezing temperatures, plague will remain alive for months, but is killed in 15 minutes at 75 degrees Celsius. It remains viable for some time in dry sputum, flea feces, and buried bodies but is killed within hours by sunlight. The mortality rate for bubonic is 25-50 percent and for pneumonic is 90-100 percent.

Q-Fever

Coxiella burnetii in known as Q-Fever and is a bacteria-like organism. It can be found in the milk of infected cows, sheep, and goats and in the dust from cattle barns. Inhalation of the dust or drinking of the raw milk infects man. Its symptoms are those of the flu and its incubation period is 2-3 weeks. It is rarely fatal.

VIRUSES

Smallpox

Smallpox is caused by the variola virus and was declared eradicated in 1980 by the World Health Organization. The United States stopped vaccinating its citizens in the early 1980s and its military in 1989. The incubation period averages 12 days. Its symptoms are malaise, fever, rigors, vomiting, headache, backache and a rash. This rash progresses to lesions all over the body. The mortality rate is 30 percent in unvaccinated victims and 3 percent in those vaccinated.

Venezuelan Equine Encephalitis (VEE)

Venezuelan Equine Encephalitis is a virus which occurs in horses, mules and donkeys and is transmitted by mosquitoes, but can also be transmitted by aerosols in weapon form. It has flu-like symptoms with a 1-5 day incubation period. It is rarely fatal and is killed by heat and disinfectants. It was weaponized by the U.S. in the 1950s and 1960s before the germ warfare treaty.

TOXINS:

Staphylococcal Enterotoxin (SEB)

Staphylococcal Enterotoxin causes food poisoning when ingested but also can be inhaled in an aerosolized form. It has an incubation of 4-6 hours. Its symptoms vary depending on whether it is ingested or inhaled. If ingested, its symptoms are cramps, vomiting and diarrhea, and if inhaled, cough, and high fever with chills. Recovery from inhalation is 1-2 weeks, while recovery from ingestion can be 6-8 hours. Lasting health problems and death is rare.

Figure 4–9 Decontamination class.

Botulinum

Botulinum is one of the most toxic substances known to mankind. It can be contracted by eating food from improperly canned food goods and poor food handling, but it can also be aerosolized in weapon form and inhaled. Symptoms can show up 24-72 hours after either ingestion or inhalation of the toxin and can include dry mouth and throat, blurred vision and trouble talking or understanding speech.

The mortality rate is around 60 percent but an antitoxin is available. Iraq has been known to have filled and deployed over 100 munitions with botulinum according to some military sources.

Ricin

Ricin is produced from the castor bean and is easily made. It can be ingested or inhaled, and the incubation is 18-24 hours if inhaled and 24-72 hours if ingested; death can occur 36-72 hours from onset of symptoms. The mortality rate is 100 percent.

CHEMICAL AGENTS

Chemical warfare has been around for centuries. Germany developed chemicals for use in war before WWI, but it took the Gulf War and the Tokyo subway incident to bring it home to first responders. All chemical agents are liquids or gases, which liquefy at low temperatures and are dispersed as a gas or aerosol. Because they are a gas or aerosol, weather and terrain influence their effectiveness.

NERVE AGENTS

Nerve gas or nerve agents were developed before WWII in Germany but were not used. There are four nerve agents—Tabun (GA), Sarin (GB), Soman (GD), and VX— and all have similar properties. They are a close relative to the organophosphate pesticides, which are widely used in the U.S., as well as other countries. They affect the central nervous system and thus are called nerve agents. They are liquids, which volatilize easily and will evaporate, so they are nonpersistent except for VX. The agent VX is a newer, recently developed agent with a low volatility, so it is more of a liquid hazard than the others.

Figure 4–10 WWI chemical bombs.

All are inhalation hazardous in a gas or aerosol form, and VX is also a skin contact hazard. All are fast acting, but antidotes are available. Although they are liquids at room temperature, they are weaponized as vapor or tiny droplets. The military refers to them as gases to get the soldiers in the mind set to wear gas masks in an attack. Yelling "droplet, droplet" just does not have the effect of yelling "gas, gas." This is the reason soldiers are issued gas masks. First responders should also have this mind set.

When pure, G-agents are colorless and odorless, but may have a slight fruity odor. VX is odorless, but may have a slight yellow color and the smell of sulfur.

Symptoms can include drooling, a running nose, difficulty in breathing, pinpointed pupils in the eyes, nausea, twitching, unconsciousness and death.

CHOKING AGENTS

There are two choking agents: Phosgene (CG) and Chlorine (Cl). Firefighters should already be familiar with these two chemicals. Phosgene is a by-product of heat and some old time fire-extinguishing agents such as carbon tetrachloride and Chlorine is encountered at transportation accidents, swimming pools, and water treatment sites. Because both chemicals are widely used in industry, terrorists have easy access to them. At low temperatures and high pressures, both are liquid but become gases quickly when released. This makes them nonpersistent, and they will quickly dissipate. Because they are heavier than air, their vapors will sink to low spots, such as trenches and basements.

Figure 4-11 New York City Police at WMD exercise.

Their symptoms are usually immediate in high concentrations, with coughing, burning eyes, choking, and tightness in the chest.

Phosgene has an odor of newly mown hay, although most people are not familiar with that smell, while **Chlorine** has a household bleach odor.

BLOOD AGENTS

Hydrogen Cyanide (AC) is one of the two blood agents. As a liquid, it is called Hydrocyanic Acid, and as a gas, Hydrogen Cyanide. Both have the chemical formula HCN. HCN is the gas used in the gas chamber, and it is also produced by the burning of vinyl plastics. It is used in many industrial processes, and over 300,000 tons per year are used in the U.S. alone.

It is just slightly lighter than air so the vapor cloud will rise.

Cyanogen Chloride (CK) is the other blood agent, and like HCN is a gas but it is heavier than air.

Both gases smell like bitter almonds and both kill by not allowing the oxygen in the blood to be released to the tissues of the body. They are inhalation threats and work immediately, causing victims to appear flushed with reddish skin and lips for light-skinned people and blue for dark-skinned people. The symptoms are very similar to Carbon monoxide poisoning victims. Victims will gasp for air, fall unconscious, and die with 6-8 minutes after exposure to high concentrations.

BLISTER AGENTS (VESICANTS)

There are three basic groups of blister agents. The first one is called *mustard agent* and it is a family of three: H, HD, and HN. We will treat all three as one, as their properties are much the same. They are light yellow to brown in color and are oily liquids. They have a strong fish or garlic odor. H and HD freeze at 57 degrees F and all are volatile at room temperature. H and HD were produced in the early 1880s and used in WWI. Although it produced the majority of the chemical causalities, it caused only 5 percent of the fatalities. Egypt used it against Yemen in 1960, and Iraq used it against Iran in 1980. The U.S. is currently destroying its stockpile of mustard agent.

The effect of mustard depends upon its concentration, with eye pain and damage occuring within a few hours, and respiratory damage and blisters in 2-24 hours. It can be fatal in large concentrations.

Symptoms include reddening of skin, blisters, eye reddening and pain, coughing, and airway pain and damage. With H there is no immediate pain.

Lewisite (L) is the second of the blister agents and was discovered in 1918, too late to be used in WWI. It causes immediate pain to the eyes and respiratory system. Its vapors like H are heavier than air. It is sometimes mixed with H to lower the freezing temperature of H. Its symptoms are similar to H with eye pain, coughing, and blisters after 6-24 hours.

Phosgene Oxime (CX) is the third blister agent. It has an irritating odor and is a solid below 95 degrees F, but can give off enough vapors to produce victims. Some metals cause it to decompose.

CX causes immediate eye and throat pain, but does not form blisters. It will cause pain upon contact with skin, and the area will become blanched within 30 seconds, surrounded by a red ring.

Figure 4–12 FDNY Rescue 4 suits up in level a chemical suits.

■ CONCLUSION

Terrorists using weapons of mass destruction are a real threat, but there is no need to panic. Now that the first responders are aware of the threat, they can plan and train for it just as they have for all the other dangers they face in their duties. An excellent resource book is *Terrorism Handbook for Operational Responders* by Armando Bevelacqua and Richard Stilp, Delmar Learning.

■ REFERENCES

* Medical Management of Biological Casualties Handbook
  U.S. Army Medical Research Institute of Infectious Diseases, 2nd Edition, Aug. 1996

* Medical Management of Chemical Casualties Handbook
  U.S. Army Medical Research Institute, 2nd Edition, 1995

▪ A ▪
COLOR INSERT CAPTIONS

Figure 1   Nonelectrical blasting caps. Note the brightly colored plastic connectors and the quarter used to show size.

Figure 2   Electrical blasting caps. Note the two wires extending out the back that are shunted together. Note the cardboard tube containing a blasting cap.

Figure 3   Safety fuse. The safety fuse has a black core, and the detonating cord has a white core.

Figure 4   Spools of detonating cord. Different colors indicate different strengths.

Figure 5   Bags of ammonium nitrate fertilizer. It will not burn or explode in an unconfined and pure state.

Figure 6   A 50-pound bag of blasting agent, ammonium nitrate and a fuel.

Figure 7   Kinepax, a bag of ammonium nitrate and a tube of nitromethane that gives it a pinkish color and strong odor. Note the quarter for size.

Figure 8   Booster explosive for blasting agent. A blasting cap is inserted into the booster.

Figure 9   Different types of boosters.

Figure 10   Stick of dynamite.

Figure 11   Box of dynamite.

Figure 12   Sawdust filler dynamite.

Figure 13   Clay filler dynamite.

Figure 14   Types of dynamites.

Figure 15   Tube of slurry explosive.

Figure 16   Stick of dynamite and tube of slurry on a box of dynamite.

Figure 17   Smokeless powders and black powders.

Figure 18   Types of fuzes. Shown are a mixture of bomb and projectile fuzes.

Figure 19   Types of bomb nose fuzes. (Left to right, top) AV-1 D/U (Russian), M 145 (U.S.), M 126A1 (U.S.), ADV (Russian); (bottom) M 103PD (U.S.), M 904 with booster (U.S.), M 904 without booster, M 173 (U.S.).

Figure 20   M 904 nose fuze in a Mk 82 500-pound bomb. Note the arming wire and window in the fuze. Red in the window means armed; green means unarmed but still explosive.

Figure 21   Nose fuzes. (Left to right) T-45 booster cup of explosives that fits into the bomb with the fuzes fitting into it, 960 Mk 1 electrical fuze (British), AN-M 23A1 (old designation) or FMU-7 A/B (new designation) (U.S.); (bottom) M 103A5 (U.S.). Note the quarter for size.

Figure 22   Tail fuzes. (Left to right, top) M 905 (U.S.), FMU-26/B (U.S.), FMU-72 (U.S.), FMU-39/B (U.S.), AVU-E (Russian), AVU-B (Russian); (bottom) M 44 arming vane (U.S.), M 190 (U.S.). Note the quarter for size.

Figure 23   ATU-35 drive assembly for the tail fuze on the side of a Ballute tail fin on a Mk 82 500-pound general purpose bomb.

Figure 24   M 188 VT proximity fuze on an M 41 20-pound fragmentation bomb (U.S.). Note the spiral wound steel "wire" for fragmentation.

Figure 25   M 168 VT proximity fuze (U.S.).

Figure 26   M 532 VT proximity fuze (U.S.).

Figure 27   MK 339 "rosette" nose fuze used on cluster bombs such as the Mk 20 Rockeye.

Figure 28   Fuzes. (Left to right, top) M 429 rocket fuze (U.S.), AV-1 modified (Russian), AM-A bomblet fuze (Russian), M-12 (Russian), M 524 (U.S.), DKZB (Russian), MDL-VDM (Czechoslovakian); (bottom) MDL-28/21 B35 (French), M 190 (U.S.). Note the quarter for size.

Figure 29   Hand grenades. (Left to right, top) Chinese stick grenade, North Vietnamese stick grenade; (middle) M 26 (U.S.), M 33 (U.S.), MK 2 (U.S.), S.R.C.M. Model 35 Italian grenade; (bottom) RKG-3M (Russian). Note the quarter for size.

Figure 30   M 2 "pineapple" grenade (U.S.), with spoon and O-shaped pull ring removed.

Figure 31   Smoke and chemical grenades. (Left to right) M 34 WP smoke, M 18 smoke, M 7A3 riot control CS gas, M 25 riot control (all are U.S.).

Figure 32   Egyptian rifle grenades. (Top) HOSAM type 3 with spring; (bottom) HOSAM type 1.

Figure 33   Rifle grenades. (Clockwise from left) M 7 high-explosive antitank (HEAT) (U.S.), HEAT (Hungarian PGK), 40-mm riot control CS (U.S.), AZ-58K-100 (Germany), 40 mm (U.S.), 81-mm HEAT (Polish), 40 mm (U.S.), NR 235 A3 57-mm silent mortar (U.S.), 40 mm (U.S.), HEAT (U.S.), M 23 (U.S.).

Figure 34   RPGs. (Top) OG-7 (Russian); (bottom) M 57 (Yugoslavian).

Figure 35   Various types of projectiles. Note the different colors.

Figure 36   Artillery shells up to 16 inches in diameter. Note that U.S. shells are normally olive drab in color unless they are chemical shells, which are gray with different-color bands. Russian shells are normally gray. (See color chart for details.) Note the rotating bands at the base of the shells. These shells have never been fired.

Figure 37   A 155-mm projectile with an M 532 VT fuze (U.S.). Note (1) the yellow band indicating high explosive filler and (2) the rifling on the rotating band at the base indicating that this round has been fired and that the fuze is armed.

Figure 38   Recoilless rifle shell. The rotating band is near the top rather than the base.

Figure 39   Sabot projectile cutaway view. It is solid metal with no explosive components.

Figure 40	Flechettes. These tiny solid metal darts are contained in both artillery and rockets called bee hive rounds and are indicated by yellow diamond markings. Note the quarter for size.
Figure 41	Mortar shells. (Top) M 30A3, 81-mm white phosphorus (WP) illumination mortar; (bottom) Stokes mortar (both are U.S.).
Figure 42	Mortar shells. (Left to right) M 58 120-mm (Yugoslavian), M 69 90 mm (U.S.), MDLBK 881 (Russian), 82 mm (Russian), M 374 81 mm (U.S.), 81 mm (Israel); (top) 60 mm Type 31 (Russian).
Figure 43	Japanese knee mortar.
Figure 44	Shoulder-fired Russian rocket-propelled grenade (RPG).
Figure 45	Light antitank weapons (LAWs). (Left to right, top) RPG-7, MDL RPG-9, RPG-18, RPG-22 (all Russian); (bottom) U.S. M 72 LAW launcher. U.S. rockets look similar to RPGs.
Figure 46	A 2.75-inch rocket with pod and pylon that attaches to the aircraft. The warhead is olive drab with a yellow band and the rocket motor is white with a brown band.
Figure 47	Rear view of two types of 2.75-inch rocket pods.
Figure 48	Dragon antitank missile (U.S.). These missiles are wire guided to the target.
Figure 49	Rear view of a Dragon antitank missile showing wire spool and wire.
Figure 50	Stinger antiaircraft missile (U.S.). These missiles are shoulder fired and seek the heat from the aircraft exhaust. Each is about 4 feet long and 3 inches in diameter.
Figure 51	Stinger launcher and container. It is a fire-and-forget missile, and the launcher is discarded. (Photo courtesy of Hughes Aircraft.)
Figure 52	Tube-launched, optically tracked, wire-guided (TOW) missile pod (outboard) and 2.75-inch rocket pod (inboard).
Figure 53	TOW missile family with different warheads. These missiles are wire guided to target. (Photo courtesy of Hughes Aircraft.)
Figure 54	AIM-9 Sidewinder heat-seeking missile (outboard) and AIM-7 Sparrow radar-guided missile (inboard). Both are air-to-air missiles. They may also be white or olive drab with a yellow band for the warhead and a brown band for the rocket motor.
Figure 55	Different missiles, including a TOW at the very top.
Figure 56	U.S. antipersonnel (APERS) mines. (Top) M 18A1 Claymore; (middle, left to right) M 86 PDM, M 74 ADAM, M 75 RAAMS; (bottom) M 14. Note the quarter for size.
Figure 57	U.S. Claymore mine. This cutaway view shows the ball bearings that serve as shrapnel.
Figure 58	U.S. BLU-92/B Gator APERS mine. Trip wires come out of the four holes on the top. The mine detonates when disturbed. The BLU-91/B is similar in appearance to this mine but is an antitank mine.
Figure 59	U.S. M 7A2 APERS mine.
Figure 60	Chinese Type 72 APERS mine. It has an antidisturbance fuze.
Figure 61	Russian PMN-2 APERS mine.
Figure 62	Russian PMN APERS mine. Note the quarter for size.
Figure 63	Italian VS-50 APERS mine. Note the quarter for size.
Figure 64	APERS mines. (Clockwise from left) M 56 aerial dispursed mine, AP (French), MDL 59 "Inkstand" (French), POZM-2M (Russian); (center) PRB-M35 (Belgium).

Figure 65   Bounding-type mines nicknamed "Bouncing Bettys." When
             tripped, an explosive charge shoots up about 3 feet and then
             explodes. (Clockwise from left) V-69 Valmara (Italian), M 16A1
             (U.S.), M 2A1 (U.S.), OZM-3 (Russian).
Figure 66   Antitank mines. These mines are larger than APERS mines. (Left
             to right, top) TC-6 (Italian), TC-2.4 (Italian), HCT-2 (Italian), FMK-3
             (Argentinian), VS-HCT (Italian); (bottom) TC-3.6 (Italian), VS 1-6
             (Italian), VC-2 (Italian), VSAR 50 (Italian).
Figure 67   U.S. submunitions (BLUs). (Left to right, top) M 83 "butterfly bomb,"
             Dragon tooth mine, BLU-91/B Gator mine; (bottom)
             BLU-24/B, M 39 grenade, BLU-54/B mine. Note the quarter for size.
Figure 68   U.S. BLU-54/B cutaway view showing the ball bearings used for
             shrapnel. Note the quarter for size.
Figure 69   U.S. BLU-3. Its fins fold around the bomblet before release.
Figure 70   Russian submunitions. (Left to right, top) AD-1, RTAB, ZAB incen-
             diary, PTAB; (bottom) PFM-1. Note the quarter for size.
Figure 71   (Top) Russian PFM-1; (bottom) U.S. Dragon tooth. Note the quar-
             ter for size.
Figure 72   U.S. antitank BLUs. (Clockwise from left) M 42 DP, BLU-7, BLU-97,
             BLU-91 Gator, M 23. Note the quarter for size.
Figure 73   Antitank BLUs. (Top) HB 876 (British); (bottom, left to right)
             M 118 (U.S.), BLU-77 (U.S.), BLG-66 (French), information not
             available on the next three. Note the quarter for size.
Figure 74   Mk 20 Rockeye cluster bomb unit (CBU) and bomblets. (Left to
             right) BLU-7, M 118, BLU-77, BLU-4/B, BLU-3, BLU-17, ISCB-1 area
             denial cluster weapon, BLU-49A/B.
Figure 75   CBU-59/B cutaway view showing the BLU-77/Bs inside (U.S.).
Figure 76   Aircraft ejection seat rocket motor. An ejection seat has many
             explosive items.
Figure 77   Pylon and cartridges. The cartridge is installed if the nut is show-
             ing, and it is not installed if the holder is sticking out. Both are
             shown here.
Figure 78   Rear of pylon cartridge.
Figure 79   Front of pylon cartridge.
Figure 80   (Left) AN-76 incendiary bomb; (middle) AN-57 500-pound general
             purpose bomb; (right) AN-30 100-pound general-purpose bomb.
Figure 81   Mk 82 500-pound general-purpose bomb with "snake eye"
             retarding fins and an M 904 nose fuze.
Figure 82   GBU-15 smart bomb. Smart bombs are regular iron bombs with a
             guidance kit.
Figure 83   Practice bombs. (Top) Mk 106; (bottom) an M 76 (old designation)
             or BDU-33 (new designation). Small practice bombs are relatively
             safe although some have a small spotting charge in the nose.
Figure 84   A 20-pound M 41 fragmentation bomb with a 0A1 impact fuze.
Figure 85   MC-1 chemical bomb. See color chart for details.
Figure 86   TNT.
Figure 87   C-3 plastic explosive. It is light yellow in color and can be molded.
Figure 88   C-4 explosive. It is white in color and can be molded like putty.
Figure 89   Sheet explosive. It is dark green in color.
Figure 90   M 60 fuse lighter. The M 60 is a match in a plastic waterproof
             tube. The safety fuse is inserted into the end.

Figure 91   M 50A3 incendiary bomb with cluster adapter. These bombs were used in World War II to bomb cities.

Figure 92   World War II bombs. An AN-M 30A1 100-pound general-purpose bomb minus its tail fin is shown on the left; two AN- 47A3 incendiary bombs are shown in the middle and on the right.

Figure 93   BLU-1C/D napalm bomb. The BLU-52 chemical bomb is similar in appearance. Napalm is gasoline and soap.

Figure 94   Napalm bomb fuze. AN-M 23A3 is the old designation, and FMU-7A/D is the new designation. This fuze contains WP, which ignites when exposed to air.

Figure 95   A fuel air explosive (FAE) bomb. These bombs consist of a flammable gas such as propane that creates a flammable mist that ignites seconds after impact.

Figure 96   Chaff and flare dispenser used to confuse enemy radar and heat-seeking missiles. The same dispenser ejects both.

Figure 97   Aircraft flare used as a decoy for heat-seeking missiles.

Figure 98   Chaff package. It has a small explosive charge to expel ribbons of aluminum foil to fool radar.

Figure 99   Aircraft flare used for illumination.

Figure 100   Type of aircraft illumination flare.

Figure 101   M 127A1 handheld flare.

Figure 102   (Left) M 115A2 artillery simulator; (right) M 116A1 hand grenade simulator.

◼ B ◼

COLOR CODING

The following color code charts should be used only as a general guide. As stated earlier, the color and markings should not be trusted for the following reasons:

1. Different countries have different color codes.
2. Colors may have been painted over, especially on war souvenirs and training items.
3. Discoloring may occur due to sun bleaching, rusting, and such.
4. Booby-trapped items or mines may be colored to blend in with the surroundings. For example, a mine may be one color for jungle use and another for desert use.
5. Color coding has been revised and changed several times over the years.

The main change in color coding with U.S. and NATO ordnance has been to chemical weapons. The body of the ordnance is still gray, but the number and color of the stripes or bands and the markings have undergone several revisions, depending on the chemical filler. This color change has led to confusion for even trained personnel. Exact color coding is beyond the scope of this book and is not necessary for emergency personnel to know. The following charts are presented without elaboration or explanation.

Table B-1 Former Soviet Union Color Codes for Bombs

Nose Band	Body Band	Type
Green	Blue	Fragmentation
Green	Blue and green	Fragmentation and chemical
Orange		Semiarmor piercing
Blue		Armor piercing
Red	Blue	Incendiary
	Red	Incendiary dispenser
Red	Green	Persistent chemical
Green	Green	Nonpersistent chemical
White	White	Parachute flare
Blue	Black	Rocket assisted
Red	White	Practice

See chart in the color insert.

Table B-2 Former Soviet Union Color Codes for Projectiles

Color Codes (Bands)	Type
Red	Incendiary
Blue	Concrete piercing
Black	Smoke
White	Illumination
Yellow	Ball shrapnel
Khaki	Bar shrapnel
One green band	Nonpersistent agent
Two green bands	Persistent agent

See chart in the color insert.

Table B-3 Former Soviet Union Markings for Chemical Munitions

Marking	Chemical Filler
P-4	White phosphorus
P-5	Mustard agent
PC	Lewisite agent
P-10	Phosgene agent
P-15	Adamsite (DM)
TP	Thermite

See chart in the color insert.

CHECKLIST

☐ Turn off all electromagnetic radiation (EMR)–transmitting devices such as two-way radios, TV mobile units, and radar within 1,000 feet.

☐ Clear and control the area as you would for any hazardous material incident. The size of the area depends on the size of the item, type of item, terrain, and such.

☐ Notify the explosive ordnance disposal (EOD) personnel responsible for your area.

☐ Have fire and emergency medical service (EMS) units stage outside the control area.

☐ Do *not* approach or disturb the item.

☐ Reduce the blasting and shrapnel effect by shielding and venting.

At this point, you have done all you can do. Leave the disarming to the EOD personnel.

■ D ■

GLOSSARY

ADAM — Area denial mine is a land mine designed to prevent access into an area.

ADAMSITE (DM) — A vomiting agent added to tear gas to prevent the use of a gas mask.

AGM — Air to ground missile fired from a helicopter or fixed wing aircraft at a ground target.

AIM — Air intercept missile, used to shoot down another aircraft.

Alpha Particle — A heavy, positively-charged particle that is the result of a radioactive material decaying.

AMAT — Antimaterial ordnance which is designed to destroy objects rather than people.

Ammo — A slang term for ammunition or ordnance.

Ammunition — Frequently called ammo, ammunition ranges from as small as 7.62 mm up to 16-inch shells.

ANFO — A mixture of ammonium nitrate fertilizer and fuel oil and used as a blasting agent.

ANTHRAX — A bacteria spore found worldwide in livestock.

AP — AP stands for armor piercing ordnance.

APERS — Antipersonnel ordnance designed to kill people.

AT — Antitank ordnance designed to destroy tanks and armored vehicles.

ATF — The U.S. government's Bureau of Alcohol, Tobacco and Firearms, which is responsible for regulating explosives and investigating bombings.

BDU — Bomb dummy unit is an inert training munition containing no explosives.

Beta Particle — A high-energy electron from a radioactive material, which is the result of the decay of a radioactive material.

Black and Smokeless Powders — These are common over-the-counter low explosives, used by gun enthusiasts to reload ammunition or used in old powder-loading guns. Unconfined, they will deflagrate (burn intensely), but will detonate when confined. They are most often used in homemade bombs.

Blasting Agents A combination of a fuel and oxidizers such as fuel oil and ammonium nitrate fertilizer (NH4NO3), sometimes known as nitrocarbonitrate or ammonium nitrate fuel oil (ANFO).

Blasting Caps Small cylinders containing a small amount of a powerful explosive, used to initiate the main charge. They are set off by either electrical charge, shock or fire.

BLISTER AGENTS Chemicals which affect the skin surface.

BLOOD AGENTS Chemicals which affect the exchange of oxygen in the blood.

BLU Bomb live unit is also called a submunition or bomblet. These are small bombs within a container called a cluster bomb unit (CBU).

Booby Trap Any bomb disguised as something other than a bomb.

Boosters Explosives in which blasting caps are placed to increase the power of the initiating charge.

CAD Cartridge activated devices are small explosive devices which serve a variety of uses on an aircraft.

CBRNE An acronym which stands for chemical, biological, radiological, nuclear, explosive to describe weapons of mass destruction.

CBU Cluster bomb unit which is a munition containing many smaller bombs.

Chemical Ordnance These can be as simple as a riot grenade or smoke bomb or as lethal as Sarin nerve agent. U.S. chemical weapons are typically gray with colored bands.

Cholera Cholera is any of several gastrointestinal diseases, marked by severe diarrhea.

CS A riot control agent commonly called tear gas.

Detonating Cord Similar in appearance to a safety fuse, it comes in various colors on spools holding up to 2,000 feet of cord. All detonating cord has a white core of a powerful explosive. This cord is normally used to set off multiple charges.

DP Dual purpose, like a munition which is both antipersonnel and antitank. ·

Dynamites There are two basic types of dynamites—nitroglycerin and ammonium. Dynamite comes in a variety of sizes, shapes and colors. Cartridges, or sticks, range typically from 1 1/8 to 3 inches in diameter and from 8 to 24 inches in length, but can be larger. Cartridges range in color from light brown to orange-red in fiberboard or waxed-paper sticks, and are typically shipped in 50-pound cardboard boxes.

EOD   Explosive ordnance disposal, which is the technical name for bomb squads, i.e., EOD units. The slang name for a bomb technician is a crab, taken from the shape of the EOD badge or insignia.

EMR   Electromagnetic radiation, which is energy in a wave form, such as radio, TV, and light.

EMS   Emergency medical service is a term given to the first responders who render medical aid at the scene of emergencies.

FAE   Fuel air explosives; they are relatively new, so much of the information is unavailable.

FUDS   Formerly used defense sites or old military installations. Many contain unexploded ordnance (UXO)

Fuzes   Fuzes are used to initiate the explosive in any warhead, whether a bomb, projectile, rocket, or any piece of ordnance. They are different from fuses.

Gamma   An electromagnetic wave similar to X-rays.

GP   General purpose such as a bomb with both concussion and fragmentation effect.

Grenades   One of the most common types of military ordnance. Many are found in the homes of ex-military personnel.

Half-life   The time it takes for one half of a radioactive material to decay.

HE   HE stands for high explosive.

HEAT   High explosive antitank.

HEP   High explosive plastic.

ICBM   A rather old term for an intercontinental ballistic missiles such as the MX missile.

IED   Improvised explosive device, a fancy name for a homemade bomb.

Incendiaries   Devices that burn rather than explode. They are classified by the U.S. Department of Transportation as Class 1.4 explosives.

LAW   Light antitank weapon.

MDT   Mobile data terminals are small computers used in police cars, fire trucks or EMS units. These tiny computers put out EMR just as a two-way radio does.

Mines   Explosives designed to (1) destroy an enemy in place or (2) prevent or deny the enemy entry into an area.

Mm   Stands for millimeter, a unit of metric measurement.

Mortar Shells   They look much like projectile shells except that they have no brass casing or rotating band.

NATO	The North Atlantic Treaty Organization.
NBC	Nuclear, biological, chemical when referring to weapons of mass destruction.
Nerve Agents	Chemicals which affect the central nervous system.
PD	Point detonated refers to a type of fuze that goes off by impact with the target.
Plague	A contagious disease caused by bacteria.
RAAMS	A type of remote antiarmor mine system.
RDD	A radiological dispersal device is often called a dirty bomb and is a conventional explosive with radioactive material attached.
Round	A slang term for a projectile.
RPG	Rocket propelled grenade, these are the weapons of choice by the terrorists in Iraq.
RSP	Rendering safe procedure, a method of disarming or defusing ordnance. RSPs are usually classified.
Safety Fuse	A time-delay device much like a firecracker fuse.
SAM	Surface to air missiles are designed to shoot down aircraft.
Sarin (GB)	Sarin is an organophosphate nerve agent.
Shell	A slang term for a projectile.
Simulators	Large firecrackers that simulate battle noise and effects.
Slurries or Gels	These are among the newer explosives. Slurries are liquid explosives, and gels are semi-liquid explosives. Either can be poured and are stored in large sausage-shaped tubes.
Soman (GD)	Soman is an organophosphate nerve agent.
SOP	Standard operating procedures are preplanned instructions and guides for a situation.
Submunitions	These can be mines or small bomblets. They are called bomb live units (BLUs) and are air-dropped from aircraft, helicopters, or projectiles.
SUU	Suspended underwing unit is a dispenser for cluster bombs and other small ordnance.
Tabun (GA)	Tabun is an organophosphate nerve agent.
TNT	Trinitrotoluene is a high explosive chemical by which all other explosives are measured.
TOW	A tube launched, optically tracked, wire guided missile fired either on the ground or in the air.
Toxins	Toxic substances produced naturally by animal, plant, or microbe.
U.N.	The United Nations.